jacaranda *plus*

Next generation teaching and learning

eBook *plus*

Access all formats of your online Jacaranda resources in three easy steps!

To access your resources:

1. go to **www.jacplus.com.au**
2. log in to your existing account, or create a new account
3. enter your unique registration code(s).

Note
- Only one JacPLUS account is required to register all your Jacaranda digital products.
- By registering the code(s) within your JacPLUS bookshelf, you are agreeing to purchase the resource(s). Please view the terms and conditions when registering.

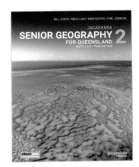

REGISTRATION CODE

Electronic versions of this title are available online; these include eBookPLUS and PDFs. Your unique registration codes for this title are:

9ACF7UAZHKC
9KAJR5VSV5Q
9FXMVQXGH7S

Each code above provides access for one user to the eBookPLUS and PDFs.

NEED HELP?

If you would like to discuss specific digital licensing options or request digital trials, or if you require any other assistance, email **support@jacplus.com.au** or telephone 1800 JAC PLUS (1800 522 7587).

A Wiley Brand

Jacaranda Senior Geography 2 for Queensland Units 3 & 4 Third Edition

JACARANDA
SENIOR GEOGRAPHY 2
FOR QUEENSLAND
UNITS 3 & 4 | THIRD EDITION

JACARANDA
SENIOR GEOGRAPHY 2
FOR QUEENSLAND
UNITS 3 & 4 | THIRD EDITION

BILL DODD
MICK LAW
IAIN MEYER
PHIL O'BRIEN

jacaranda
A Wiley Brand

Third edition published 2020 by
John Wiley & Sons Australia, Ltd
42 McDougall Street, Milton, Qld 4064

First edition published 2001
Second edition published 2009

Typeset in 11/14 pt TimesLTStd

© Bill Dodd, Mick Law, Iain Meyer, Phil O'Brien 2001, 2009, 2020

The moral rights of the authors have been asserted.

ISBN: 978-0-7303-6904-2

Reproduction and communication for educational purposes
The Australian *Copyright Act 1968* (the Act) allows a maximum of one chapter or 10% of the pages of this work, whichever is the greater, to be reproduced and/or communicated by any educational institution for its educational purposes provided that the educational institution (or the body that administers it) has given a remuneration notice to Copyright Agency Limited (CAL).

Reproduction and communication for other purposes
Except as permitted under the Act (for example, a fair dealing for the purposes of study, research, criticism or review), no part of this book may be reproduced, stored in a retrieval system, communicated or transmitted in any form or by any means without prior written permission. All inquiries should be made to the publisher.

Trademarks
Jacaranda, the JacPLUS logo, the learnON, assessON and studyON logos, Wiley and the Wiley logo, and any related trade dress are trademarks or registered trademarks of John Wiley & Sons Inc. and/or its affiliates in the United States, Australia and in other countries, and may not be used without written permission. All other trademarks are the property of their respective owners.

Front and back cover images: © Arco Images GmbH / Alamy Stock Photo

Cartography by Spatial Vision Innovations Pty Ltd, Melbourne, www.spatialvision.com.au, MAPgraphics Pty Ltd Brisbane and the Wiley Art Studio

Typeset in India by diacriTech

Printed in Singapore by
Markono Print Media Pte Ltd

All activities have been written with the safety of both teacher and student in mind. Some, however, involve physical activity or the use of equipment or tools. **All due care should be taken when performing such activities**. Neither the publisher nor the authors can accept responsibility for any injury that may be sustained when completing activities described in this textbook.

A catalogue record for this book is available from the National Library of Australia

10 9 8 7 6 5 4 3 2

CONTENTS

About this resource vi
About eBookPLUS viii
Acknowledgements ix

UNIT 3 RESPONDING TO LAND COVER TRANSFORMATIONS — 1

1 Land cover transformations and climate change — 3
- **1.1** Overview — 3
- **1.2** What is land cover and its distribution? — 4
- **1.3** The Earth's physical systems — 9
- **1.4** Global climate systems — 12
- **1.5** Changes to land cover — 21
- **1.6** Anthropogenic activity and how it has transformed land cover — 27
- **1.7** Anthropogenic activity and global warming — 55
- **1.8** Review — 74

2 Responding to local land cover transformations — 77
- **2.1** Overview — 77
- **2.2** Processes resulting in land cover change — 78
- **2.3** The spatial pattern of land cover change — 86
- **2.4** The effects of land cover change — 99
- **2.5** Connections between people and physical systems — 103
- **2.6** How can we mitigate our negative impacts on the land? — 106
- **2.7** Fieldwork in the local area — 110
- **2.8** Review — 115

UNIT 4 MANAGING POPULATION CHANGE — 117

3 Population challenges in Australia — 119
- **3.1** Overview — 119
- **3.2** Demographic concepts — 120
- **3.3** Changes in population — 130
- **3.4** Factors affecting population change — 136
- **3.5** Population patterns and trends — 147
- **3.6** Local area population patterns — 158
- **3.7** Implications for people and places of demographic change — 170
- **3.8** Review — 182

4 Global population change — 185
- **4.1** Overview — 185
- **4.2** Global patterns of population growth — 186
- **4.3** An ageing world — 202
- **4.4** Patterns of changing population distribution and density — 210
- **4.5** People on the move: international and internal migration — 216
- **4.6** Review — 245

Glossary — 247
Index — 254

ABOUT THIS RESOURCE

Jacaranda Senior Geography for Queensland 2 (Units 3 & 4) Third Edition is tailored to address the intent and structure of the new Senior Geography syllabus, and to inspire students' sense of geographical curiosity. The *Jacaranda Senior Geography for Queensland* series provides easy-to-follow text and is supported by a bank of resources for both teachers and students. At Jacaranda we believe that every student should experience success and build confidence, while those who want to be challenged are supported as they progress to more difficult concepts and questions.

Building a sense of inquiry with strong geographical knowledge and skills

Chapter openers begin with key questions to encourage students to begin forming questions about a topic.

Every chapter includes a range of case studies and examples to build a strong understanding of the impacts of geographical processes and patterns.

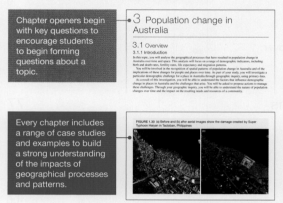

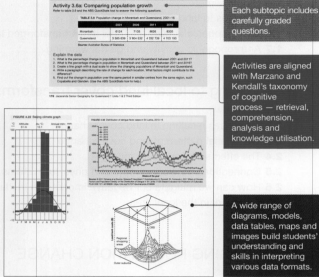

Each subtopic includes carefully graded questions.

Activities are aligned with Marzano and Kendall's taxonomy of cognitive process — retrieval, comprehension, analysis and knowledge utilisation.

A wide range of diagrams, models, data tables, maps and images build students' understanding and skills in interpreting various data formats.

Preparing students for internal and external assessment success

Two complete, unseen data analysis activities with short- and extended-response questions are provided for two of the four chapters. These can be modified to suit your class with one available in the teacher eGuidePLUS for teachers only, and one available for student self-assessment in the eBookPLUS. These tasks have been carefully designed to build students' confidence and ability with unseen data analysis tasks, and in constructing written responses to a wide range of data types. Sample responses are also provided for each activity.

An extensive glossary of terms is provided in print and as a hover-over feature in the eBookPLUS.

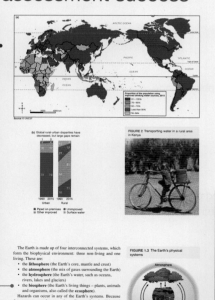

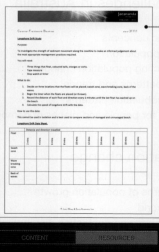

Extensive support for both teachers and students is provided for both Fieldwork and Data reports in two of the four chapters. These include an annotated coastal Fieldwork booklet and scaffold, and Data report resources and scaffold. These are available in the eBookPLUS. For teachers, a Data report and Fieldwork Teacher advice document provide extra resources and advice, available only in the eGuidePLUS.

Free sample answers are provided for chapter activities, enabling students to get help where they need it, whether at home or in the classroom — help at the point of learning is critical.

eBookPLUS features

Weblinks: Direct access to an extensive range of additional data sources and information through links placed at the point of learning.

Topic PDF: Downloadable PDF of the entire chapter of the print text.

Interactivities and Video eLessons: Bonus, step-by-step SkillBuilder instructional videos and multimedia activities consolidate students' core geographical skills. These are placed at the point of learning to enhance understanding and establish strong connections between knowledge and practical skills.

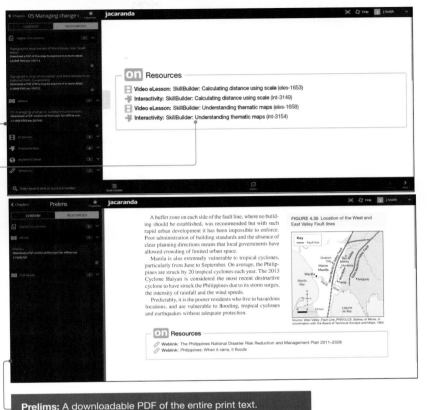

Prelims: A downloadable PDF of the entire print text.

Additional resources for teachers available in the eGuidePLUS

Teacher digital documents: Access to two quarantined assessment activities. Short answer and extended response tasks are provided with sample answers. Activities and sample answers are downloadable in Word format to allow teachers to customise as they need.

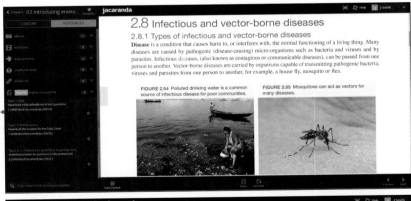

Teacher digital documents: Teaching notes and work programs are provided to assist with classroom planning.

Teacher weblinks: Teachers have access to an extensive list of case study resources, each with annotations to assist with course mapping and planning.

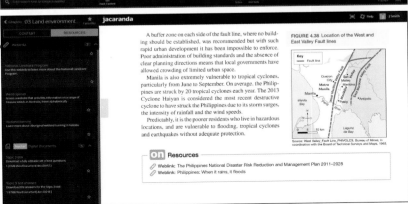

ABOUT THIS RESOURCE vii

About eBookPLUS

jacaranda plus

This book features eBookPLUS: an electronic version of the entire textbook and supporting digital resources. It is available for you online at the JacarandaPLUS website (**www.jacplus.com.au**).

Using JacarandaPLUS
To access your eBookPLUS resources, simply log on to **www.jacplus.com.au** using your existing JacarandaPLUS login and enter the registration code. If you are new to JacarandaPLUS, follow the three easy steps below.

Step 1. Create a user account
The first time you use the JacarandaPLUS system, you will need to create a user account. Go to the JacarandaPLUS home page (**www.jacplus.com.au**), click on the button to create a new account and follow the instructions on screen. You can then use your nominated email address and password to log in to the JacarandaPLUS system.

Step 2. Enter your registration code
Once you have logged in, enter your unique registration code for this book, which is printed on the inside front cover of your textbook. The title of your textbook will appear in your bookshelf. Click on the link to open your eBookPLUS.

Step 3. Access your eBookPLUS resources
Your eBookPLUS and supporting resources are provided in a chapter-by-chapter format. Simply select the desired chapter from the table of contents. Digital resources are accessed within each chapter via the resources tab.

Once you have created your account, you can use the same email address and password in the future to register any JacarandaPLUS titles you own.

Using eBookPLUS references
eBookPLUS logos are used throughout the printed books to inform you that a digital resource is available to complement the content you are studying.

Searchlight IDs (e.g. **INT-0001**) give you instant access to digital resources. Once you are logged in, simply enter the Searchlight ID for that resource and it will open immediately.

Minimum requirements
JacarandaPLUS requires you to use a supported internet browser and version, otherwise you will not be able to access your resources or view all features and upgrades. The complete list of JacPLUS minimum system requirements can be found at **http://jacplus.desk.com**.

Troubleshooting
- Go to **www.jacplus.com.au** and click on the Help link.
- Visit the JacarandaPLUS Support Centre at **http://jacplus.desk.com** to access a range of step-by-step user guides, ask questions or search for information.
- Contact John Wiley & Sons Australia, Ltd. Email: support@jacplus.com.au Phone: 1800 JAC PLUS (1800 522 7587)

ACKNOWLEDGEMENTS

The authors and publisher would like to thank the following copyright holders, organisations and individuals for their assistance and for permission to reproduce copyright material in this book.

Images
• .id: **164, 165, 169 (bottom), 170** • AAP: **104 (top)**/GRAEME MCCRABB / AAP; **179 (bottom)**/Richard Durham / AAP; **233**/Reuters Graphic / AAP • AAP Newswire: **219 (top)**/EPA / United Nation Relief and Works Agency • Above Photography: **162**/© Mike Swaine • Alamy Australia Pty Ltd: **119**/CulturalEyes - AusGS2; **178 (bottom)**/Bill Bachman; **215 (bottom)**/imageBROKER • Alamy Stock Photo: **221**/Islandstock; **224 (bottom)**/Shawshots; **226 (bottom)**/Andrew Melbourne; **242**/adrian arbib • © 2019 Aussie Towns: **167 (top)** • Australian Bureau of Agricultural and Resource Economics and Sciences, Forests of Australia 2013: **112 (top)** • Australian Bureau of Statistics: **121, 126, 133, 136, 138, 139 (top), 141 (top), 142, 148, 149, 150, 152, 153, 156, 157, 158, 159, 163 (bottom), 168, 168–9, 171, 171–2, 172, 176 (top), 178 (top), 179 (top), 180 (top)** • Australian Foreign Affairs: **154 (top)**/George Megalogenis • Australian Institute of Health and Welfare: **137** • Barcelona Field Studies Centre: **199** • BBC: **56**/EC Joint Research Centre/PBL Netherlands Environmental Assessment Agency; **64**/Verisk Maplecroft • Bill Dodd: **24 (top left, top right), 26, 30, 31 (top left, top right), 32, 50** • Bureau of Meteorology: **19, 66 (top), 68 (top), 69, 98, 100** • Central Intelligence Agency: **124 (top)** • Christopher Frederick Jones: **178 (centre)** • Climate Action Tracker: **59** • Copyright Clearance Center: **201–2**/International Monetary Fund 2017; **225 (bottom)**/Thomas Lee Philpott; **232 (bottom)**/Christian Dustmann and Tommaso Frattini, The Fiscal Effects of Immigration to the UK, The economic journal : the quarterly journal of the Royal Economic Society, 124:580, November 2014. • Creative Commons: **162 (bottom)**/J Bar / Wikimedia Commons; **192**/Thomas Robert Malthus. Mezzotint by John Linnell, 1834. Credit: Wellcome Collection; **226 (top left), 228, 232 (top), 239**/The Office for National Statistics; **235**/United Nations Population Division Technical Paper No. 2013/1 - Cross-national comparisons of internal migration: An update on global patterns and trends; United Nations Development Programme Human Development Index 2013, https://creativecommons.org/licenses/by/3.0/igo/; **240**/© Crown Copyright and Office for National Statistics 2014 • CSIRO: **99** • Department of Environment and Energy: **52** • Dr Philip O'Brien: **173 (bottom), 174 (top)** • EAAFP: **53 (top left), 54 (bottom)** • Elsevier: **54 (top)** • European Commission, Eurostat: **218 (top), 223 (top), 230–1** • FAO: **29 (bottom)** • Geoscience Australia: **6, 103, 163 (top), 167 (bottom)** • Getty Images: **42 (bottom)**/Vittorio Ricci – Italy • Gravity Consulting: **139 (bottom)** • Green Collar Group: **104 (bottom)** • Japan, Statistics Bureau: **214** • John Wiley & Sons Australia: **130**/Source: © 2019 Population Reference Bureau, United Nations, Our World in Data; **180 (bottom)**/Adapted from Queensland Health; **186 (bottom)**/Created based on information from Worldometers and United Nations; **203 (bottom)**/Based on data from World Health Organization; **205 (bottom)**/Sources: World Population Review, Our World In Data & World Bank Group; **210**/Based on statistics from United Nations • John Wiley & Sons Inc: **197 (top)** • MAPgraphics: **22, 36, 41, 45** • Michael Morrish: **204 (top), 205 (top left)** • Mongabay: **39 (bottom), 40 (top)** • Museum of London: **223 (bottom)**/© Estate of Roger Mayne • NASA: **57, 85**; **5**/Reto Stöckli, NASA Earth Observatory; **7**/Boston University and NASA GSFC • National Archives of Australia: **144 (top left)** • National Snow and Ice Data Center: **70** • Natural Earth: **219 (bottom)** • Newspix: **166 (top)**/Lyndon Mechielsen • NOAA: **15, 21, 66 (bottom)** • NSW State Archives and Records: **140 (top left)** • Our World in Data: **9 (top), 120, 200**; **89**/Max Roser and Hannah Ritchie; **186 (top), 187**/Max Roser and Esteban Ortiz-Ospina • Out of Copyright: **38**; **225**/Park & Burgess, 1925, The City: Suggestions for Investigation of Human Behavior in the Urban Environment". • Oz Aerial Photography: **162** • Paul Hayler: **205 (top right)** • Peter Veth: **108** • Population Pyramid: **129, 134, 135, 190, 194, 207** • Population Reference Bureau: **123** • Public Domain: **60**/Climate Emergency Institute; **144 (top right)**/National Archives of Australia: A12111, 1/1960/16/54; **206**/Wikipedia; **212 (top), 229, 236 (top), 243 (right)**/Made with Natural Earth. Free vector

and raster map data @ naturalearthdata.com • Queensland Government: **81, 90**/© The State of Queensland Queensland Treasury 2019; **92**/© The State of Queensland Queensland Treasury 2018; **93 (top left, top right), 94, 95, 96, 109, 116, 161**/© State of Queensland, 2018; **151, 154–5**/© The State of Queensland Queensland Treasury 2017; **162, 166 (bottom)**/The State of Queensland, Department of Infrastructure, Local Government and Planning; **173 (top)**/© The State of Queensland Department of Education 2018; **176 (bottom), 179 (centre)**/© The State of Queensland Department of Education 2019 • Robert Murdie: **224 (top)** • Shutterstock: **1**/Warren Chan; **3**/Sk Hasan Ali; **4**/Harvepino; **12, 14**/Designua; **13 (top)**/udaix; **17**/ValentinaKru; **20**/Labrynthe; **23**/Tutti Frutti; **24 (bottom right)**/a2l; **29 (top left), 86 (bottom)**/AustralianCamera; **24 (bottom left)**/Brisbane; **29 (top centre)**/Kaiskynet Studio; **29 (top right)**/SSSCCC; **31 (top centre)**/GUDKOV ANDREY; **31 (centre)**/Juliann; **31 (bottom left)**/Ondrej Prosicky; **34 (top)**/Quick Shot; **34 (bottom)**/Tarcisio Schnaider; **40 (bottom right)**/guentermanaus; **40 (bottom left)**/Rainer Lesniewski; **43 (left)**/Marius Dobilas; **43 (right)**/Nina B; **44 (bottom)**/Denis Burdin; **44 (top)**/Michael Wick; **47 (bottom)**/Gervasio S. _ Eureka_89; **48 (bottom)**/BOULENGER Xavier; **49, 71**/Peter Hermes Furian; **53 (top left)**/aDam Wildlife; **53 (bottom right)**/HelloRF Zcool; **53 (bottom left)**/hxdbzxy; **63**/livoeian; **72 (top)**/Matyas Rehak; **72 (bottom)**/Salvacampillo; **79**/SkyReefPhoto; **83 (bottom)**/Alizada Studios; **87**/John Lee; **101**/berm_teerawat; **110**/stefanolunardi; **114**/Joyseulay; **115**/Ivan Chudakov; **119**/Visun Khankasem; **140 (top right)**/Gorodenkoff; **140 (bottom right)**/Joachim Heng; **147**/FabianGame; **175 (top)**/Pawel Papis; **175 (bottom)**/Songsook; **184**/Christian Bertrand; **217 (top)**/Nicolas Economou; **230 (top)**/Ivsanmas; **238 (bottom)**/Reservoir Dots; **243 (left)**/Ryan Rodrick Beiler • Skyepics: **111, 162** • Spatial Vision: **13 (bottom)**; **191**/Redrawn by Spatial Vision based on information from The International Monetary Fund; **193 (top)**/Central Intelligence Agency; **211**/The Earth Institute Columbia University / United Nations; **217 (bottom)**/Eurasian Research Institute, UNHCR, Eurostat; **218 (bottom)**/Redrawn by Spatial Vision based on the information from IHS Conflict Monitor and UNHCR • Springer: **20**/Campos, Camila & Horn, Myriel. 2018. Figure 3, The Physical System of the Arctic Ocean and Subarctic Seas in a Changing Climate: Proceedings of the 2017 conference for YOUng MARine RESearchers in Kiel, Germany. 10.1007/978-3-319-93284-2_3. • Springfield City Group: **177** • State Library of NSW: **140 (bottom left)**/Mitchell Library • Statistics Indonesia: **241** • The Conversation: **122**/Alasdair Rae • The World Bank: **93 (bottom), 124 (bottom), 125, 131 (top), 141 (bottom)** • Trust for London: **238 (top)** • UCL Geomatics: **25**/© ESA Climate Change Initiative - Land Cover project led by UCLouvain 2017 • United Nations: **131 (bottom), 132**/World Population Prospects 2017 by United Nations / Population Division. Copyright © 2017 United Nations; **188, 189 (top)**/World Population Prospects 2017 by United Nations / Population Division. Copyright © 2017 United Nations; **193 (bottom)**/Demographic Components of Future Population Growth: 2015 Revision by United Nations. Copyright © 2015 United Nations; **201 (top)**/ World Population Ageing Report 2015 by United Nations / Population Division. Copyright © 2015 United Nations; **204 (bottom)**/World Population Prospects: The 2010 Revision, Volume I: Comprehensive Tables by United Nations / Population Division. Copyright © 2010 United Nations • United Nations Statistics Division: **208 (bottom)** • University of Alaska Fairbanks: **55**/Katey Walter Anthony • University of Manchester: **226 (top right), 227** • University of Maryland: **37**/Hansen/UMD/Google/USGS/NASA • US Census Bureau: **195 (bottom)**; **127, 128**/United Nations, Department of Economic and Social Affairs, Population Division 2017. World Population Prospects: The 2017 Revision, custom data acquired via website • Visualizing Economics: **197 (bottom)**/Catherine Mulbrandon • Weatherzone: **47 (top)** • West End State School: **174 (bottom left, bottom right)**/© The State of Queensland 2018 • World Bank Group: **198, 203 (top), 209, 222** • Yirrganydji Gurbana Aboriginal Corporation: **106**

Text

• Geography 2019 v1.1 – General Senior Syllabus © State of Queensland (Queensland Curriculum & Assessment Authority) 2019; Headings, various glossary terms/definitions • African Development Bank Group: **44** • Public Domain: **64**/Bill Dodd, Spectrum: Geographical Perspectives on People and their Environment, p208, Jacaranda, 1994, based on data provided by CSIRO • Great Barrier Reef Marine Park Authority: **65** • Queensland Farmers Federation: **88** • Queensland Government: **95**

Every effort has been made to trace the ownership of copyright material. Information that will enable the publisher to rectify any error or omission in subsequent reprints will be welcome. In such cases, please contact the Permissions Section of John Wiley & Sons Australia, Ltd.

Acknowledgements 2019 V5.1 – General Source Syllabus © State of Queensland (Queensland Curriculum & Assessment Authority) 2019, reading, various classroom definitions • Africa: Development Bank of Africa, 14 • Public Domain, 64 Jill Dopp, Specimens, Geographical Perspectives on People and their Environment, p22r, Ascemide, 1965, Based on data provided by CSIRO • Great Barrier Reef Marine Park Authority, 68 • Queensland Floods – Federation X • Queensland Government, 95

Every effort has been made to trace the ownership of copyright material. Information that will enable the publisher to rectify any error or omission in subsequent reprints will be welcome. In such cases, please contact the Permissions Section of John Wiley & Sons, Australia, Ltd.

UNIT 3
RESPONDING TO LAND COVER TRANSFORMATIONS

The Earth is changing. Both natural and anthropogenic geographical processes are affecting the land cover of Earth and leading to climate change. For people at a global, local and regional scale, these changes create impacts and challenges that must be managed.

In this unit you will study the impact of land cover changes at a global and local scale. You will look closely at your own local area through fieldwork and study the challenges of a land or water management issue.

CHAPTER 1   Land cover transformations and climate change (Unit 3, Topic 1) 03

CHAPTER 2   Responding to local land cover transformations (Unit 3, Topic 2) 77

UNIT 3
RESPONDING TO LAND COVER TRANSFORMATIONS

The Earth is changing, both natural and anthropogenic geographical processes are changing the land cover of Earth and leading to climate change. For people at a global, local and regional scale, these changes create impacts and challenges that must be managed.

In this Unit you will study the impact of land cover changes like a pick of land where you will look closely at your local area through fieldwork and study the changes to the land or water management here.

CHAPTER 1 Land cover transformations and climate change Unit 3, Topic 1

CHAPTER 2 Responding to local land cover transformations Unit 3, Topic 2

1 Land cover transformations and climate change

1.1 Overview

1.1.1 Introduction

As the Earth's population continues to grow, there has never been a more demanding time for governments and communities to provide food, water and shelter for so many. With a current global population of almost 8 billion, and increasing by 80 million per year (about 120 000 each day), demand for living space and basic **resources** has probably reached an optimum level. However, the pressure to provide these essential needs, as well as supply industrial and technological goods, and energy and infrastructure comes with a price tag. This cost is an escalation in the level and intensity of exploitation of the Earth's existing land, water and mineral resources, but these resources are finite. The challenge for your generation and for the future is to create an acceptable balance between demand and sustainable supply.

In addition to over-exploitation of resources, many areas are now forced to deal with the threat of **climate change**. Recently, there have been more severe and destructive weather events, longer and hotter periods, a higher frequency of catastrophic wildfires as well as **inundation** of coastal wetlands and low-lying islands due to ocean warming and sea level rise.

In this topic you will explore some of the physical and human geographical processes that have resulted in changes to surface land areas, and the subsequent spatial patterns evident today. You will also examine how these changes may be linked to climate change and the implications for present and future generations.

FIGURE 1.1 Climate change makes extreme weather events such as floods more likely.

1.1.2 Key questions

- What is land cover and its distribution?
- What processes connect the Earth's physical systems and affect land cover?
- What are the different types of land cover (vegetation biomes, biogeographic areas, anthropogenic biomes)?
- How does population growth, an increase in affluence and technology impact upon land cover?
- How do human activities like settlements, croplands, rangelands and forestry transform land cover surfaces?
- How do these transformations impact upon the Earth's systems?
- What are global climatic systems?
- What is climate change? How does it impact on land cover types?
- What are the implications of climate change on people and the environment and how might people best respond to them?

1.2 What is land cover and its distribution?

1.2.1 Continents and oceans

The Earth is covered by land and water. Land covers about 30 per cent of the surface and is made up of seven large landmasses called **continents**, as well as many large and small islands (see figure 1.4). The continents are Africa, Asia, Europe, North America, South America, Australia and Antarctica. Examples of large islands are Sri Lanka, Borneo, New Guinea, Madagascar and New Zealand, while inhabited small islands and **archipelagos** include Indonesia, Japan and the United Kingdom. Although a continent, Australia is often linked to Oceania, a region that includes the south Pacific Islands. Asia has the largest land area and population, and Africa is second largest.

FIGURE 1.2 Only 30 per cent of the Earth is covered by land.

Between the continents are very large basins of water called **oceans**, which contain an estimated volume of about 1.35 billion km^3 of water. This huge expanse covers 70 per cent of the surface and is divided into four major oceans — the Pacific, Atlantic, Indian and Arctic Oceans. The largest three (Pacific, Atlantic and Indian) join in the far south, creating another ocean called the Southern Ocean, which was only officially recognised as an ocean in 2002. The Southern Ocean describes seawaters south of 60-degree latitude and flows as a huge clockwise current around Antarctica. Oceans also contain smaller sections called seas (e.g. Mediterranean Sea and Caribbean Sea), as well as large gulfs (e.g. Persian Gulf and Gulf of Mexico), and bays (e.g. Hudson Bay and Bay of Bengal).

Because of their vast area, the combined oceans absorb a huge amount of the sun's heat. Ocean water takes longer to heat than continental land, but once heated it retains warmth longer and moderates air temperatures at the surface. Oceans are able to transfer some of this heat into the atmosphere and other parts of the world by powerful convectional currents. All oceans are connected and water circulates around the entire planet, playing a key role in the exchange of heat and moisture between other physical systems — the atmosphere and lithosphere.

Despite their size, oceans have become fragile environments through exposure to human progress. Technological advances, population growth, and growing affluence have made oceans vulnerable to the effects of human activities such as fishing, shipping, coastal land clearing, marine construction and pollution. In the past, the sheer size of oceans enabled them to absorb the effects of human-induced change, but today, the intensity and speed of human change has given marine **ecosystems** little time to adjust, placing them at risk of overload and eventual collapse.

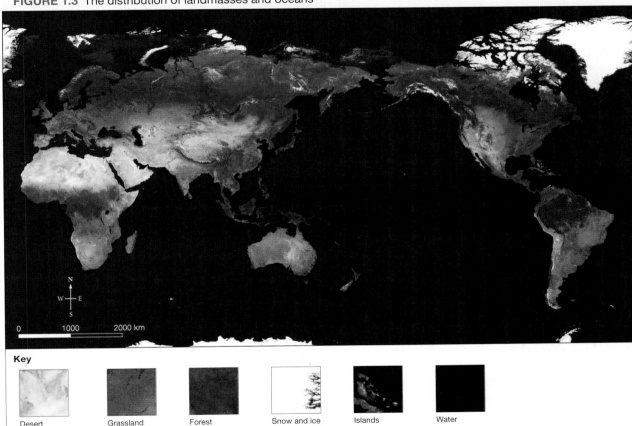

FIGURE 1.3 The distribution of landmasses and oceans

Source: Reto Stöckli, NASA Earth Observatory

1.2.2 Patterns of land cover

The term '**land cover**' is used by geographers to identify the different materials that cover the surface of the Earth (see figure 1.4). It refers to forest, grass, farmland, roads, buildings, exposed ground, lakes and water, and was first used by the plant ecologist Frederick Clements in the 1800s. *Land cover* is different from *land use*, which is a term used to explain how people use an area for economic, social or cultural purposes, such as farming, golf courses or cemeteries.

Scientists can collect accurate details of the extent of land cover using satellites. The European Space Agency (ESA) collects land cover data in 36 different classifications, including human settlements, agriculture (cultivation, grazing lands), **vegetation** (forests, grasslands, shrublands), ice sheets, areas of water and artificial surfaces. In contrast, Geoscience Australia use 16 catagories to describe land cover, listed in Table 1.1. Due to the extensive areas of land cover, they are usually mapped using **remote sensing** (satellite) techniques. Field survey work is also still important and may be required if samples need to be gathered or tested.

A problem with classifying land cover types is that different organisations may use slightly different terms or definitions. For example, according to the ESA, some areas without trees may be classified as forest cover 'if the intention is to re-plant', as in the UK. Other countries, including Norway, have areas with trees but may not be considered forest 'if the trees are not growing fast enough'.

TABLE 1.1 The 16 classes of land cover used by Geoscience Australia

Class of land cover
Urban areas
Inland waterbodies
Salt lakes
Irrigated cropping Irrigated pasture Irrigated sugar Rainfed cropping Rainfed pasture Rainfed sugar
Wetlands
Tussock grasses — Closed Tussock grasses — Open Alpine grasses — Open Hummock grasses — Closed Hummock grasses — Open Shrubs and grasses — Sparse and scattered
Shrubs — Closed Shrubs — Open
Trees — Closed Trees — Open Trees — Scattered Trees — Sparse
Built-up surface

Source: © Commonwealth of Australia Geoscience Australia 2019

FIGURE 1.4 Different types of land cover

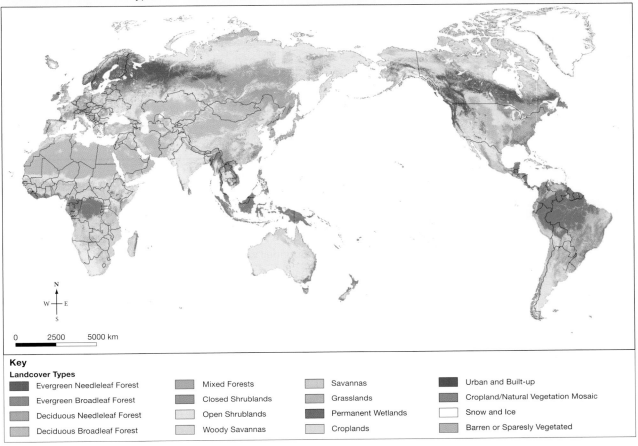

Key
Landcover Types
- Evergreen Needleleaf Forest
- Evergreen Broadleaf Forest
- Deciduous Needleleaf Forest
- Deciduous Broadleaf Forest
- Mixed Forests
- Closed Shrublands
- Open Shrublands
- Woody Savannas
- Savannas
- Grasslands
- Permanent Wetlands
- Croplands
- Urban and Built-up
- Cropland/Natural Vegetation Mosaic
- Snow and Ice
- Barren or Sparsely Vegetated

Source: Boston University and NASA GSFC

Resources

Weblink The difference between land cover and land use

Activity 1.2a: Recognising different types and patterns of land cover
Comprehend the data
1. Look through the different land cover classifications used by Geoscience Australia in table 1.1 and attempt to label them according to more commonly used terms.
2. Using the **ESA Land cover descriptions** weblink, fill in the gaps below to help you better understand the classification terms. You can also refer to the **World map of land cover types** weblinks in the Resources tab.

ESA classification	Common term example
Cropland, rainfed	Farming areas that rely on rainfall for water supply (e.g. wheat growing in Australia)
Cropland, irrigated	
	Wetland, mangroves and forests that grow in water (e.g. Kakadu wetland in NT)
Tree cover, flooded	

(continued)

(continued)

ESA classification	Common term example
Saline water	
	Snowfields, glaciers and ice caps (e.g. Antarctica)
	Rangelands used for extensive animal grazing (e.g. savanna grasslands of northern Australia)
Urban areas	

3. Match the land cover type List A with a correct activity or example in List B.

List A	List B
Closed forest	New York metropolitan area
Open forest	Boondall wetlands, near Brisbane
Grassland	Serengeti National Park, Africa
Woodland	Wheat belt of Canada
Rangeland	Brigalow woodlands of Central Queensland
Cropland	Cattle grazing on Barkly Tableland (NT)
Ice cover	Amazon Rainforest
Wetland	Eucalypt forests of Western Australia
Urban use	Greenland ice sheet

Resources

Weblinks Land cover descriptions
World map of land cover types
Land cover map

Activity 1.2b: Analysing data on a stacked area graph
Explain and analyse the data
Refer to figure 1.5 to answer the following questions.
1. Explain briefly what figure 1.5 is showing.
2. What is the dominant land cover type?
3. Estimate the percentage of land cover that is:
   (a) forestry
   (b) grassland
   (c) inland water bodies
   (d) artificial/urban surfaces.
4. Explain why the area of barren land is so large and the area of mangrove is so small.
5. Decide if these statements are true or false.
   (a) Approximately 35 per cent of land is covered by some form of tree or forest cover.
   (b) Approximately 15 per cent is regarded as barren land (mountains, deserts).
   (c) Urban and artificial areas account for approximately 2 per cent of land cover.
   (d) Most land surface is covered by trees, shrubs and grasses.

6. If global air temperatures continue to increase, which land cover is most at risk of decreasing? Why?

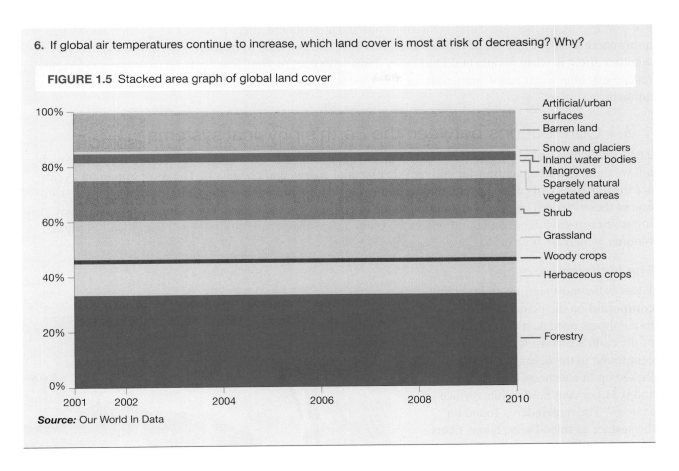

FIGURE 1.5 Stacked area graph of global land cover

Source: Our World In Data

1.3 The Earth's physical systems

The Earth comprises four physical systems — the atmosphere, the lithosphere, the hydrosphere and the biosphere. Three of these are non-living systems:
- the lithosphere (the Earth's crust and landmasses)
- the atmosphere (the mix of gases surrounding the Earth)
- the hydrosphere (water such as oceans, rivers, lakes and glaciers).

Together, the non-living systems form another system that supports life — the biosphere or ecosphere (see figure 1.6). These four systems can be divided further into smaller zones. The hydrosphere contains oceans and seas, the terrestrial parts of the lithosphere contain **biomes**, and the atmosphere is divided into layers such as the **troposphere** and stratosphere. The biosphere can be broken down into classifications such as **biogeographic areas**, ecosystems and communities.

Over time, all four systems have been largely shaped by geophysical factors such as climate, soils, vegetation and geomorphology. However, a system is a dynamic network consisting of inputs, processes and outputs. In recent times, increased levels of human activity and advances in technology have revealed that people now have more impact on these systems than ever before. If a change occurs in one part of the system, it will affect other parts of the system.

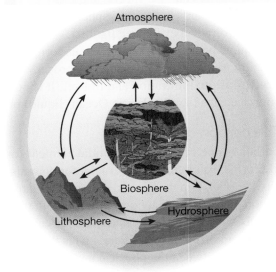

FIGURE 1.6 The Earth's systems

For example, if we increase carbon emissions into the atmosphere, there is more CO_2, which allows the atmosphere to hold more water vapour, so storms have the potential to be more violent and destructive. Another effect is when people build too many dams or extract too much water from rivers or aquifers, they reduce essential water runoff into estuarine ecosystems and wetlands. This results in the death of plants and animals.

1.3.1 Connections between the Earth's physical systems

The Earth's physical systems are connected by a complex network of pathways and loops (cycles), which allows them to exchange, transfer and recycle essential energy and chemicals. The most important cycles are the **hydrological** (water) **cycle**, the **carbon cycle** and the **nitrogen cycle**. Certain human activities in one of these systems can cause disruption to the operation of some cycles. For example, fertilisers may improve crop growth on land but if chemicals enter a river they cause pollution and the growth of **algal blooms**, which are both effects of drainage degradation.

The hydrological cycle

Water is the most significant **inorganic compound** on the planet because all living things need it. It covers most of the earth's surface with 97.2 per cent found in the oceans, 2.15 per cent locked up in ice sheets and glaciers, and 0.31 per cent held in sub-surface systems. The remainder is found on the surface as inland seas, lakes, rivers, dams and in the atmosphere. This tiny remaining percentage is probably the most life-sustaining part of the cycle.

Water exists in three main forms: liquid (ocean, rivers and lakes), solid (snow, glaciers and ice sheets) and gas (vapour or clouds in the atmosphere). The hydrological cycle (see figure 1.7) continuously circulates these various forms of water between the oceans (hydrosphere), land (lithosphere) and atmosphere through a series of processes such as evaporation, condensation, transpiration, **precipitation**, infiltration and run-off. It is possible that the water we drink today was once stored in the polar ice caps (solid), Pacific Ocean (liquid) or in the atmosphere (gas).

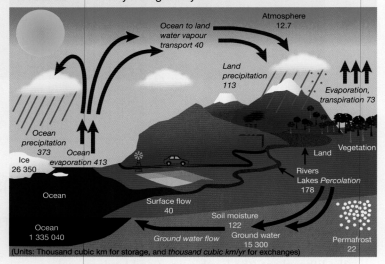

FIGURE 1.7 The hydrological cycle

The carbon cycle

Carbon is one of the most essential elements on the earth because it is the basic building block of organic life forms; its circulation is crucial for animals and plants. Carbon occurs in the atmosphere as carbon dioxide (CO_2), in the ground as coal, oil, natural gas, limestone or chalk (calcium carbonates), and is an important element of living things on both the land and in the oceans. Most living species have carbonates ($++CO_3$) in their skeletons (fish, coral) and decaying organisms produce organic carbon.

The carbon cycle (see figure 1.8) is the process of transferring, storing and exchanging some of these chemicals between the physical and living systems. For example, plants take in CO_2 from the atmosphere and store it as glucose so they can grow. An animal may eat the plant and store CO_2 in its body or exhale it back into the atmosphere. When the animal dies, it decays, and carbon is returned to the earth as organic matter.

Carbon is also exchanged between the atmosphere and the oceans. Because CO_2 is a **soluble gas**, it is taken in by marine plants such as **phytoplankton**. Just like plants on the surface, phytoplankton rely on energy from the sun to grow (**photosynthesis**). They are consumed by **zooplankton**, jellyfish, krill and baleen whales, which then become a source of carbon for fish and larger marine creatures.

The volume of CO_2 in the ocean often depends on water temperature and salinity. Cold water can hold more CO_2 than warm water so the oceans act as a huge **sink** that captures and holds large volumes of CO_2. When quantities of carbon are captured or held in storage by sinks like oceans the process is called **sequestration**. Scientists think that CO_2 is held and recycled between the atmosphere and upper ocean layers approximately every seven years, whereas it may take up to a thousand years for exchanges of carbon between the deeper ocean layers and the atmosphere.

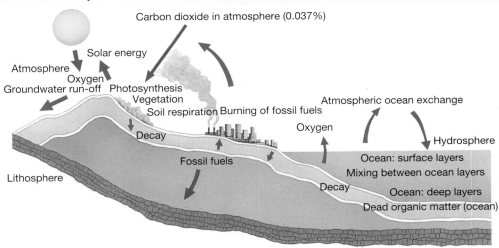

FIGURE 1.8 The carbon cycle

Forests and soils are also important carbon sinks. When a tree is growing it captures CO_2 from the atmosphere and when a tree dies or is removed, its stored CO_2 returns to the atmosphere. CO_2 stored below ground as coal, oil and gas is similarly stable. However, when humans burn these **fossil fuels**, more carbon is released into the atmosphere. Although it took millions of years to lock up this carbon, it takes only seconds to put it back in the air.

Carbon is a chemical that is not always in balance and is easily affected by human actions, such as burning or exhaust emissions. The rate at which carbon is deposited into living organisms is not the same as the rate it is returned to the Earth. Although CO_2 occupies only about 0.03 per cent, or 300 parts per million (ppm), of the atmosphere by volume, the burning of fossil fuels has contributed to a gradual increase in these levels since the **Industrial Revolution**. Readings of around 345 ppm are now common in some parts of the world.

The nitrogen cycle

Nitrogen is an important element because it is the most abundant gas in the atmosphere. Occupying about 78 per cent of the Earth's atmosphere by volume, nitrogen is also found in decaying organic matter, inorganic soil matter and marine sediment.

Despite its atmospheric abundance, organisms and plants are unable to use nitrogen in its gaseous form, except for some types of blue-green algae. The two atoms in nitrogen are held together so firmly, our bodies can only utilise it when they are separated. Animals obtain nitrogen from amino acids, the building blocks of protein. Plants absorb nitrogen through their root systems in the form of nitrates (NO_3) and ammonium salts (NH_4) when these are dissolved in water. Have you wondered why grass looks so healthy after an electrical storm? Each bolt of lightning carries adequate energy to split the nitrogen atoms (N2) in the air. They then fuse with oxygen to form nitrates and fall to the ground as a form of fertiliser.

The set of pathways and loops that transfer nitrogen between and within the physical systems is called the nitrogen cycle. While it does not readily combine with other substances, nitrogen does join with other nitrogen compounds such as nitrates (NO_3), nitrites (NO_2), or the amino group, which includes ammonia (NH_3) and ammonium (NH_4). The nitrogen cycle maintains a balance between levels in the atmosphere, the land and marine systems. However, excessive usage of nitrogen-based fertilisers can change this balance, resulting in eutrophication and algal blooms in waterways.

1.4 Global climate systems

Climate is best described as the average state of weather conditions over a long period of time, such as a month, whereas weather refers to the conditions in a much shorter space of time, such as a day or week. The global climate is created from the interaction between several systems including wind patterns, precipitation, ocean currents and the transfer of heat from the sun.

1.4.1 Wind patterns

The Earth is not heated evenly by the sun because it is round. These variations in heat create large zones where air pressure is different. In places where air is cool, air pressure is denser and falls to the surface of the Earth, creating an area of high pressure, but if air is warm, it expands and rises, forming an area of low pressure. Air moves from a high pressure to an area of low pressure to even out. These pressure differences on a large scale create a global wind system, as illustrated in figure 1.9.

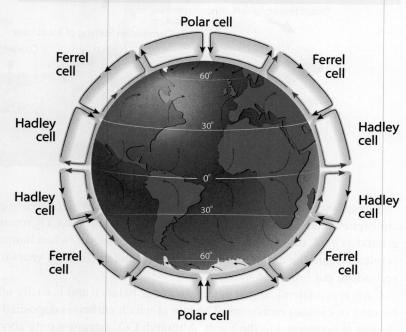

FIGURE 1.9 The global wind system

Winds are affected by the **Coriolis effect** because the Earth is rotating. This means that in the southern hemisphere winds are deflected to the left, and in the northern hemisphere they are deflected to the right. There are three major wind belts (systems) in each hemisphere — polar easterlies (from about 60–90 degree latitudes), prevailing westerlies (30–60 degrees) and **trade winds** (0–30 degrees). Around the equator, the zones between these wind belts are commonly known as the **Intertropical Convergence Zone (ITCZ)** and the subtropical convergence zones about 30 degrees north or south are called 'horse latitudes'.

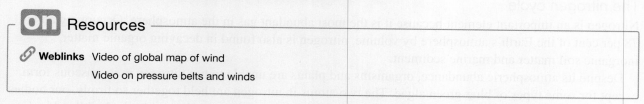

1.4.2 Heat transfer in the atmosphere

The tilt of the Earth causes physical differences between air and ocean temperatures. The Earth is tilted at an angle of 23.5 degrees, which adds to the different levels of heat absorption both on land and in the oceans as shown in figure 1.10. As well as providing a 24-hour day–night cycle, the angle of the tilt also determines our four seasons: spring, summer, autumn and winter.

The levels of solar radiation hitting the Earth are much different between summer and winter. During summer, regions directly facing the sun have large amounts of incoming solar radiation and, therefore, get quite hot. In winter, levels of solar radiation are much lower, so the land and air remain cool. Amounts of heat absorption are also affected by the angle of incidence of incoming rays. That is why temperatures are so low near the poles.

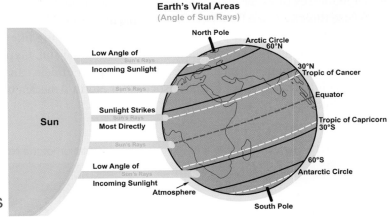

FIGURE 1.10 The angle of the Earth affects how the heat from the sun is absorbed.

1.4.3 Precipitation patterns

Due to the distribution of landmasses and the uneven surface of the Earth, availability of moisture and subsequent rainfall amounts vary considerably from place to place. As a rule, rainfall is generally highest in the tropics due to the influence of the intertropical convergence zone and instability of warm air masses. Global patterns of precipitation are closely aligned to wind patterns and heat transfer in the atmosphere. The atmospheric circulation illustrated in figure 1.9 helps to explain global patterns of precipitation. Due to increased heat transfer at the equator (see figure 1.10), the air is dense with moisture and begins to rise and condense, as shown in the **Hadley Cells** in the equatorial region (figure 1.9). These low pressure zones, where air rises, results in precipitation. This means that precipitation is greatest near the equator, as shown in figure 1.11. Natural land cover in these regions is most often tropical rainforest.

When the air in the Hadley Cell begins to sink and become warmer, evaporation and transpiration occur. These high pressure zones are found around the Tropic of Capricorn and Tropic of Cancer where precipitation is at its lowest (see figure 1.11). The world's major deserts are found in these areas.

FIGURE 1.11 Annual precipitation

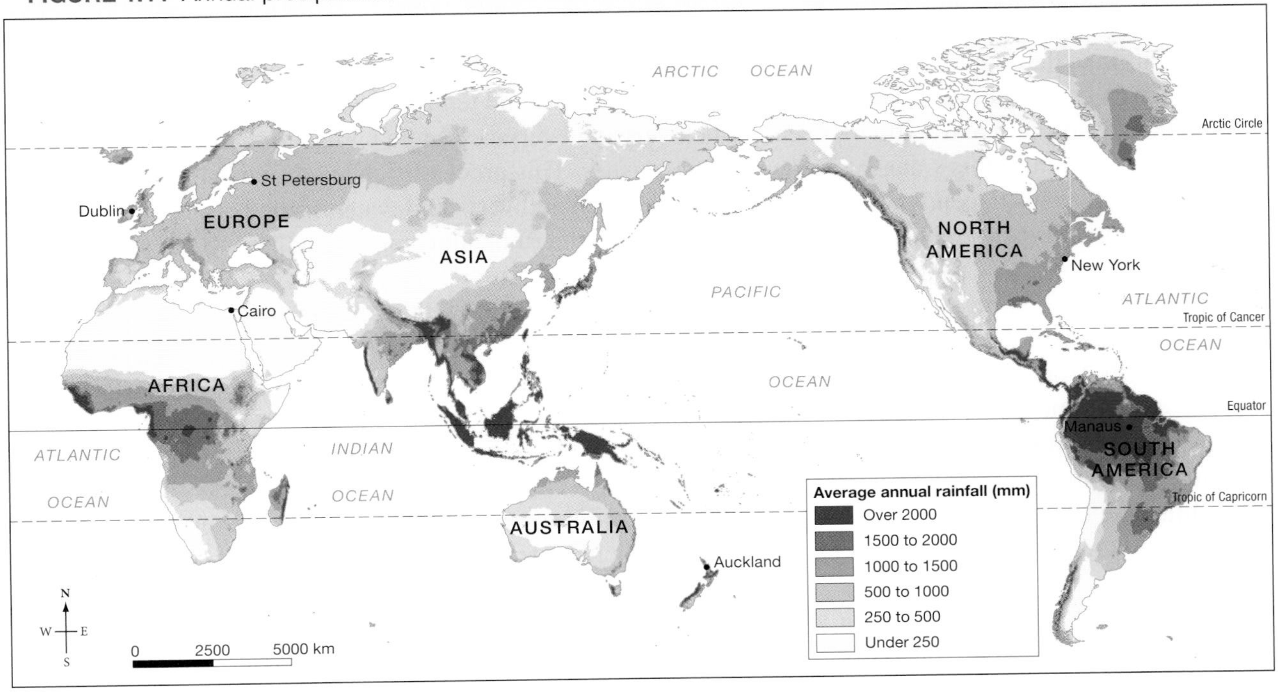

Source: © Spatial Vision

1.4.4 Ocean circulation and heat transfer

Heat energy is transferred between the equator and the poles by very large ocean currents. Ocean currents are large masses of water that circulate water flow around the oceans, either in a clockwise or anti-clockwise direction (see figure 1.12). Currents are formed by the Coriolis effect and influenced by winds, water density, tides and the shape of the ocean floor. Moving like large **gyres**, ocean currents help regulate climate by distributing uneven amounts of heat from the sun. Gyres are important controllers of climate, particularly in regions adjacent to the ocean, and their effects can be felt over large distances. For example, the warm Gulf Stream flows up the east coast of North America but still affects the climate of places in northern Europe.

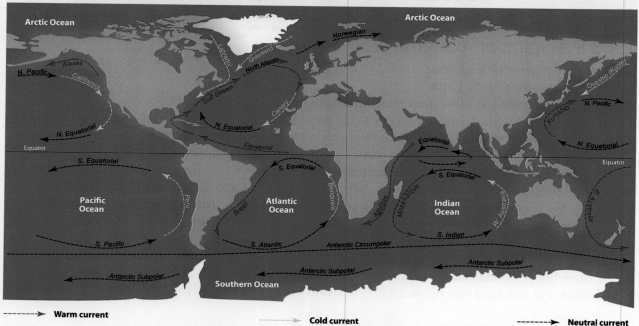

FIGURE 1.12 Surface currents and the change in circulation of the North Atlantic drift

Large currents that flow below the surface and along the sea floor are known as **thermohaline** currents, and are driven by differences in water density (mainly salinity) and temperature. Thermohaline currents form in the far northern and southern oceans when very cold surface waters plunge to the sea floor and create loops that return this water back towards the equator. This process is called **downwelling**. Because thermohaline currents depend on differences between warm and very cold water, if the oceans gain more heat and temperature differences are reduced, it is possible thermohaline currents will not be as strong or effective.

The Gulf Stream/North Atlantic drift is an important ocean current that affects the east coast of North America and parts of northern Europe. It takes warm water from the Gulf of Mexico north along the east coast of North America, crosses the Atlantic and flows towards the English and Scandinavian coasts, keeping northern ports ice-free most of the year. As the melting ice cools the water and decreases the salinity, the current sinks and then returns southward along the ocean floor. This also helps to trap CO_2 deep in the ocean. However, this current has recently shown signs of slowing down, possibly due to **global warming**. The last time occurred was approximately 1 million years ago, during the **Pleistocene Ice Age**. Even though the air around northern Europe became much colder with more snow and ice, water in the far northern oceans was warmer than the air above. There was still sufficient differences between temperature and salinity for downwelling to occur and keep a steady flow. Modern concerns are that as the oceans warm, the temperature and salinity differences will decrease. What could happen then?

> **on Resources**
>
> 🔗 **Weblink** Video on thermohaline circulation

1.4.5 The El Niño/La Niña phenomenon

El Niño and La Niña are weather events that occur when the sea surface temperature are either warmer than average or colder than average respectively. The term 'El Niño' was first used by fishers off the coast of Peru in the 17th century to describe the sudden unexplained warming of coastal waters. During 'normal' years, waters in the eastern Pacific Ocean would remain cool due to the combined effects of the icy **Humboldt Current** and upwelling from deep off-shore trenches. These two forces also brought huge quantities of marine **nutrients** to the surface, making the Peruvian coast an important fishing ground. However, the ocean would mysteriously warm up every five to nine years, causing the fish harvest to fail. At these times, the western tropical coast of South America experienced higher than usual rainfall and severe flooding, while droughts occurred across south-east Asia and eastern Australia, and North America experienced some of its coldest winter snaps.

> 'El Niño' is Spanish for 'little boy' or 'Christ child'. The Peruvians used this term because they first experienced the event around Christmas.

Scientists first thought El Niño was an occasional event that produced high rainfall over Peru and Ecuador, but data has since revealed it is part of a global event involving both the atmosphere and ocean waters. They realised there was a link through the **Walker Circulation** (Walker Cell), which is a model of air flow in the tropics in the troposphere as shown in figure 1.13.

FIGURE 1.13 The Walker Circulation

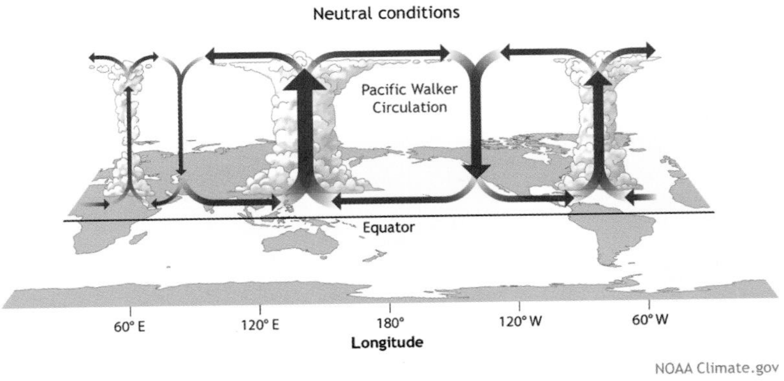

Source: NOAA Climate.gov

The Walker Circulation consists of easterly winds in the lower troposphere, an uplift of moist air over the western Pacific region, westerly winds moving through the upper troposphere and descending dry air over the eastern Pacific. It varies from year to year and sudden fluctuations trigger extreme weather conditions in many areas of the world.

The non-El Niño years are called La Niña (Spanish for 'little girl'). During this phase, warm moist air rises over Indonesia, causing cloud and rain. As air rises into the upper troposphere it is forced eastwards, cools and becomes dry. It then descends near the South American coast above the cold Humboldt Current, keeping the region dry. At the same time, trade winds at the Pacific Ocean surface drive the ocean circulation and push warm, moist air towards Indonesia and the Australian east coast. This thermally-driven atmospheric circulation generates showers or storms, and feeds energy into the **monsoon** systems and Hadley Cells. Due to the intensity of the south-easterly winds in the Australian region, the ocean level is about 40 cm higher than water levels near South America.

From observation and data, it has been established that El Niño begins when subtle changes occur to the oceanic–atmospheric circulation. Such changes include:
- if the intertropical convergence of south-east and north-east winds moves further south than usual (often shown as a monsoon trough on television weather charts)
- if surface air pressures over northern Australia begin to increase steadily, wind intensity from the Pacific onto the Australian mainland tends to weaken due to a warming of sea surface temperatures.

These changes are subtle, but their effects are enormous. Once trade winds decrease, the current that draws water from the South American coast weakens, and the bulge of water in the western Pacific (along the Australian coast) flows back towards the east. This flattens out the **thermocline** and warmer water smothers the effect of the cold upwelling along the South American coast.

The best indicator of an imminent El Niño is the strength of the **Southern Oscillation Index (SOI)**, a measure of pressure difference between the central Pacific (Tahiti) and northern Australia (Darwin). Referred to as ENSO (El Niño Southern Oscillation), the difference in surface air pressure is calculated daily and converted to an average figure, or index.

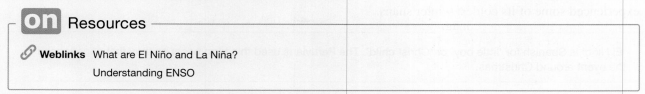

Resources

Weblinks What are El Niño and La Niña?
Understanding ENSO

1.4.6 The Earth's energy budget

Energy from the sun enters the Earth's system during the day by radiation and warms the Earth's surface. Most of the energy is absorbed by the Earth's surface but some is transferred back into the atmosphere by infrared radiation, conduction and evaporation of water (latent heat released later when water condenses). Energy also leaves the Earth during the night by infrared radiation from the atmosphere. If the atmosphere contains more CO_2, it must increase in temperature for energy balance to occur. The arrows in figure 1.14 show global average energy transfer rates in units of Watts per square metre $(W\,m^{-2})$.

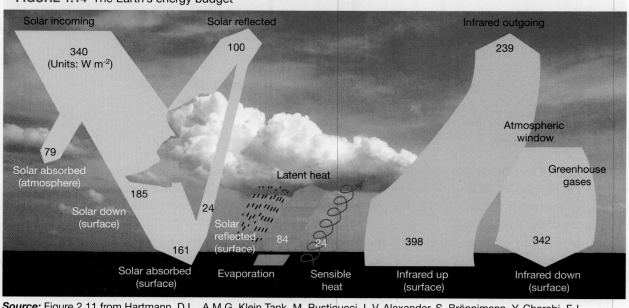

FIGURE 1.14 The Earth's energy budget

Source: Figure 2.11 from Hartmann, D.L., A.M.G. Klein Tank, M. Rusticucci, L.V. Alexander, S. Brönnimann, Y. Charabi, F.J. Dentener, E.J. Dlugokencky, D.R. Easterling, A. Kaplan, B.J. Soden, P.W. Thorne, M. Wild and P.M. Zhai, 2013: Observations: Atmosphere and Surface. In: Climate Change 2013: The Physical Science Basis. Contribution of Working Group I to the Fifth Assessment Report of the Intergovernmental Panel on Climate Change [Stocker, T.F., D. Qin, G.-K. Plattner, M.Tignor, S.K. Allen, J. Boschung, A. Nauels, Y. Xia, V. Bex and P.M. Midgley eds.]. Cambridge University Press, Cambridge, United Kingdom and New York, NY, USA.

1.4.7 The effects of surface reflectivity

Incoming solar energy is either absorbed as heat energy or reflected into space without heating the Earth. **Albedo** is a term that means the proportion of light reflected from a surface. Light coloured surfaces such as snow or ice reflect up to 95 per cent of solar energy, which means they have a high albedo. Dark areas such as rainforest, ploughed soil or ocean water absorb most of the heat and reflect only small quantities away (see figure 1.15). They have a low albedo. People alter the ability of the surface to reflect or absorb heat by changing the state of land by clearing forests, constructing buildings of concrete and glass, and laying bitumen roads.

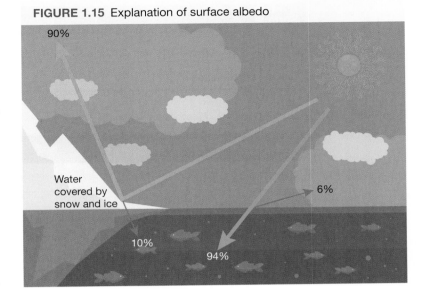

FIGURE 1.15 Explanation of surface albedo

Global warming is also altering the land surface, by affecting the rate of ice melt in many glaciers, in the Arctic and on the Antarctic ice shelf. This results in more land being exposed to sunlight so more heat is absorbed by the Earth. Seas in the far north of the world, such as the Barents Sea, north of Russia, are now ice-free most of the year. This has enormous implications for people of this area.

Activity 1.4a: Transforming data
Analyse the data and apply your knowledge
Refer to table 1.2 to answer the following questions.
1. Construct a vertical bar graph to compare levels of reflectivity. Use different colours for each of the three categories: low, medium and high.
2. Consider why there are variations in albedo between surfaces such as types of soil, snow and grasslands?
3. Describe the interconnections between land cover loss and albedo. Use generalisation to suggest the impact on climate change.

Low

TABLE 1.2 Albedo levels for common surfaces (%)

Surface	Albedo level
Sea water/lakes	6–7
Bitumen road	5–10
Dark soil	5–15
Grey soil	10–20
Rainforest	5–15
Crops	15–25

Medium

Surface	Albedo level
Desert	25–30
Savanna grasslands (dry)	25–30

High

Surface	Albedo level
Snow (fresh)	75–95
Snow (days old)	40–70
Cumulus cloud	70–90

Resources

Video eLesson Skillbuilder: Creating a simple column graph (eles-1639)
Interactivity Skillbuilder: Creating a simple column graph (int-3135)

1.4.8 What is the greenhouse effect?

Greenhouse gases are a collection of naturally occurring gases in the troposphere that allow the sun's rays through to the Earth and traps some of the heat. Scientists estimate that without protective greenhouse gases, including CO_2, the Earth's surface temperatures would be approximately $-20°C$, making the planet uninhabitable. Of the total incoming radiation from the sun, approximately 31 per cent is reflected by cloud, other air-borne particles and the Earth's surface. The remaining 69 per cent is absorbed by ozone in the stratosphere, water vapour, clouds, pollutant gases in the troposphere and the Earth's surface.

To maintain an energy balance, the Earth releases long-wave radiation equivalent to the amount of incoming short-wave radiation (69 per cent). As most of this energy comes from the surface and may be trapped by greenhouse gases in the air, surface air temperatures increase until the correct amount of energy is released. Scientists agree that because the Earth is retaining too much heat, an enhancement of the greenhouse effect that controls temperature will upset this balance and affect aspects of climate in many areas. Refer back to figure 1.14 to see how this works.

1.4.9 The Indian Ocean Dipole

The Indian Ocean Dipole (IOD) is a measure of sea surface temperatures between two places, hence the title *dipole*. A meteorologist can measure sea surface fluctuations using a western pole in the Arabian Sea (western Indian Ocean) and an eastern pole in the Indian Ocean, south of Indonesia. Alternating warm and cool ocean temperatures affect the rising and falling of atmospheric moisture, and give a clearer indication of when dry or wet spells may occur over the western mainland.

The IOD experiences both negative and positive phases. Meteorologists have concluded that when the IOD is in a negative phase, winds drive moisture towards the Australian coast and there is rain over the north-west mainland. If the phase is positive, moisture moves away from the mainland and dry spells become frequent over Western Australia. These phases may align with El Niño periods in the Pacific. According to the Bureau of Meteorology (BOM), from 1960, when reliable records of the IOD began, to 2016 there have been 11 negative IOD and 10 positive IOD events while others are neutral.

Find out more about the Indian Ocean Dipole by watching the video in the Resources tab.

FIGURE 1.16 The Indian Ocean Dipole (a) positive phase and (b) negative phase

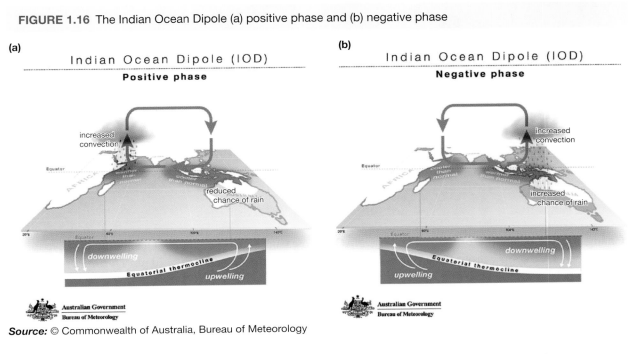

Source: © Commonwealth of Australia, Bureau of Meteorology

Resources

Weblink Understanding the Indian Ocean Dipole

1.4.10 The Arctic Oscillation and polar vortex

The far north of the world experiences considerable seasonal climatic differences in temperature, air pressure, wind and moisture from those found in the mid and low latitudes. This is because of the vast expanses of ice and snow in the region. Two phenomena that play key roles in climatic patterns and the distribution of air are the Arctic Oscillation (and its cousin North Atlantic Oscillation) and the **polar vortex** (see figure 1.17).

Arctic Oscillation

The Arctic Oscillation (AO) is part of the overall climate pattern affecting regions of the Northern Hemisphere. It is based on the pattern of counter-clockwise winds circulating around the Arctic at approximately 55°N latitude, just below the Arctic Circle. As well as circulating a strong airflow helping move air masses around the Arctic, it contains much of the icy air to the far north. There is a similar circulation in the far north of the Atlantic Ocean called the North Atlantic Oscillation (NAO). Together they control the direction and intensity of westerly winds and possible storm paths.

FIGURE 1.17 (a) The Artic Oscillation and (b) its chilling effects.

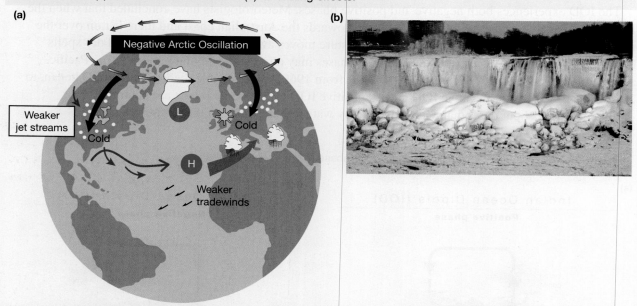

Source: Campos, Camila & Horn, Myriel. 2018. Figure 3, The Physical System of the Arctic Ocean and Subarctic Seas in a Changing Climate: Proceedings of the 2017 conference for YOUng MARine RESearchers in Kiel, Germany. 10.1007/978-3-319-93284-2_3.

Changes to the behaviour of the AO and the NAO are measured using an index. During the positive phase of the AO, the flow around the North Pole is strong with little outflow of cold air. During a negative phase, winds become weaker, thus allowing a southward migration of icy cold air and storms further south into populated countries. The level of penetration into southern countries is also affected by surface pressure cells and continental land masses (Europe, Asia and North America). The AO and NAO work in conjunction with the polar vortex, creating the undulating wave motion and jet streams (fast flowing currents of air in the upper atmosphere) that carry icy cold air into the mid-latitudes (creating freezing weather) and bringing back warm air from the south into the arctic (causing ice melt and thawing of **permafrost**).

Polar vortex

Weather in the far north is also affected by the polar vortex, a large area of swirling cold air at low pressure close to the North Pole. There is a similar vortex at the South Pole, but because Antarctica is largely uninhabited, it has a limited effect on human populations. While these vortices are a normal part of the Earth's climatology and are permanently located at the poles, they have several unique features that affect weather in high latitude regions in northern Europe, northern Asia and North America. These include:
- the vortex forms due to significant temperature differences between polar regions and the mid-latitudes
- the vortex builds up during the northern autumn and gets stronger in winter when there is no direct sunlight reaching the ice. In summer, when sunlight reaches the North Pole, the vortex weakens but remains stable, and maintains a circular flow around the North Pole.
- the vortex is situated high in the troposphere. This means that pools of cold air may be carried away from the poles by jet streams and taken to areas not accustomed to such cold snaps; this type of disruption may last for a period of four to eight weeks, causing freezing conditions in Europe, Russia, Canada and the US. This weather is lethal to communities and highly disruptive to infrastructure.

FIGURE 1.18 The polar vortex can cause abnormally cold temperatures to northern regions during winter.

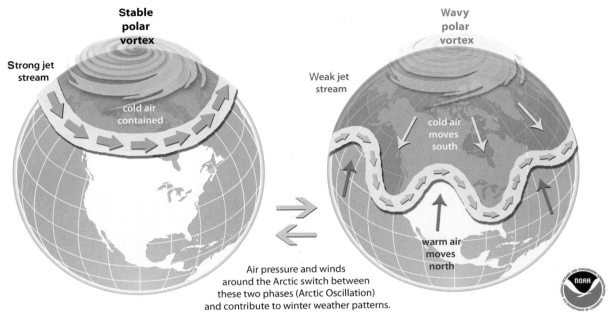

Source: National Oceanic and Atmospheric Administration

on Resources

Weblink Understanding Arctic Oscillation and the polar vortex

1.5 Changes to land cover

Natural land cover, or vegetation biomes, exist as a result of the climatic environments that have been created through the interactions between global climatic systems such as precipitation patterns, ocean circulation, heat transfer and global winds. Without any human interruptions, theses biomes would continue. An examination of changes to land cover looks at the processes that have transformed natural land cover, such as changes to land drainage, **deforestation**, intensification of agriculture, soil degradation and water degradation. These processes continue to alter natural land cover, which results in the destruction of any remaining truly 'natural' environments. This ever-changing human modification of environments has led to the development of the **anthropogenic biome**.

1.5.1 Distribution of land cover — biomes and biogeographical areas

Most terrestrial regions contain large natural landscapes where climatic conditions have been constant for thousands of years. They differ from each other by supporting diverse communities of plants (trees, shrubs and grasses) that have specifically adapted to the climates and soils of those regions. Known as biomes, they are named after their most common vegetation land cover — rainforests, dry forests, deserts, grasslands, woodlands, **tundra**, boreal (taiga) forests, mountain regions and polar regions. Biomes are also home to many animal species adapted to their vegetation. For example, a tropical grassland biome supports herbivorous (plant eating) animals such as giraffes and elephants.

FIGURE 1.19 The distribution of the Earth's major biomes

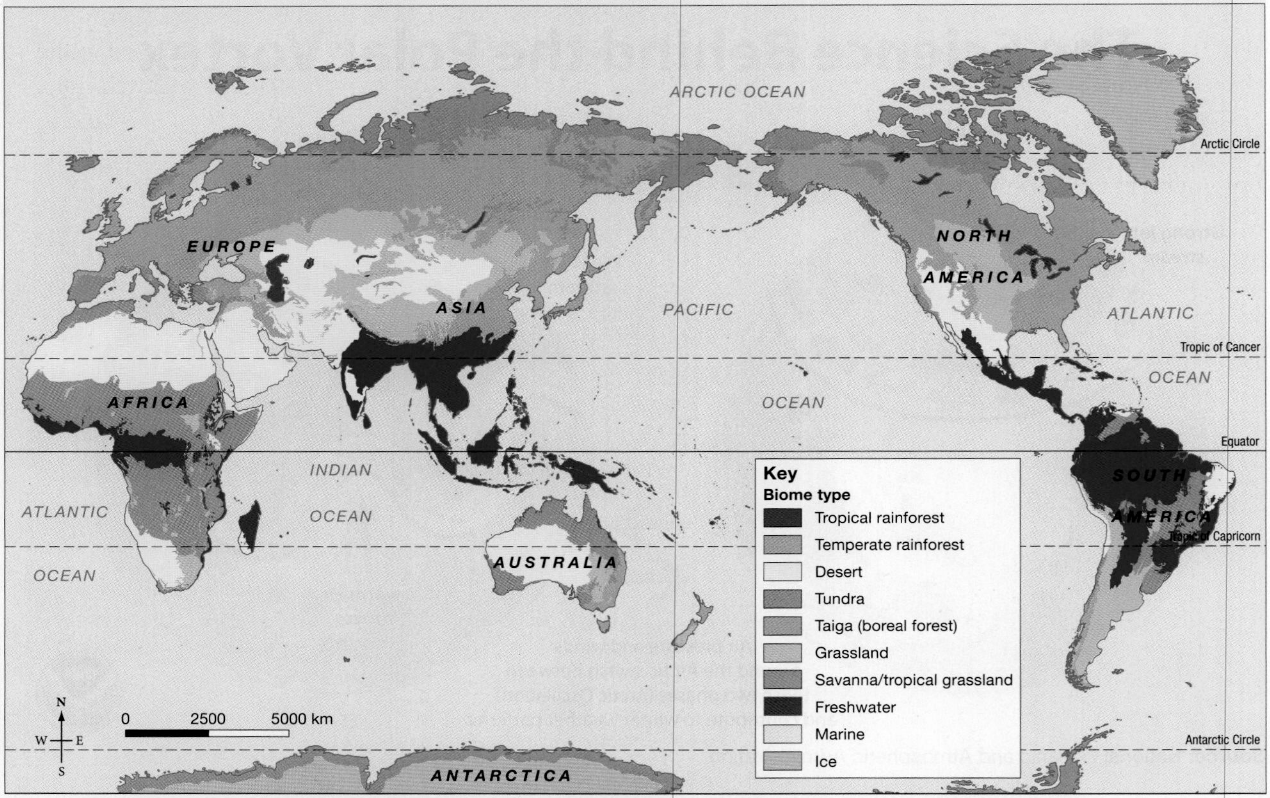

Key — Biome type: Tropical rainforest; Temperate rainforest; Desert; Tundra; Taiga (boreal forest); Grassland; Savanna/tropical grassland; Freshwater; Marine; Ice

Source: Redrawn by Spatial Vision based on the information from the Nature Conservancy and GIS Data

Biomes have their own unique landscapes and landforms, generally because of variations in climate. Each of these variations affect temperature, humidity and rainfall. The main climatic factors are:
- latitude (distance from the equator)
- distance to the sea
- elevation above sea level
- proximity to ocean currents.

Biomes also contain smaller biogeographic areas that support specific communities of plants and animals according to precise environmental conditions such as climate, soils and vegetation. Australia has 89 distinct biogeographic regions including the wet tropics of North Queensland, the Nullarbor Plain in South Australia and Western Australia, and the Australian Alps in New South Wales and Victoria. Local ecosystems, such as mangrove wetlands, fall into this category based on climate and soils.

Many biomes are favourable places for people to live because they have an abundance of natural resources such as fertile soil, water, timber and animals once hunted by humans. When human populations were small and global systems such as the atmosphere and oceans perceived to be large, people often thought that nature could take care of itself, meaning that these systems were so large they were immune from human misuse and natural catastrophe, and were capable of 'self-healing'. We know now that this is not true.

1.5.2 Factors influencing changes in land use

Increased population growth and demand for food, water and other resources have increased the pressure on the land and fertile soils. Although unoccupied grasslands and forests once existed, and a vast source of arable land lay waiting to be cultivated, people now realise this is no longer the case. Fertile arable land in particular is limited in supply due to urban expansion and human-related degradation over the past century.

Poor catchment management practices involving water removal and irrigation, dam construction, deforestation, waste disposal, overgrazing, overcultivation and chemical pollution have contributed to the degradation of land and water resources. Land cover is further degraded by occurrences such as salinity, soil acidification, soil erosion, desertification, water pollution, loss of wildlife as well as exposing land to the effects of flooding and bushfires. Many projects to sustain economic development, food security and employment have resulted in long-term changes to the Earth's biophysical systems.

Activity 1.5a: Recognising spatial patterns of land cover

Refer to figure 1.20 to answer the following questions.

Explain the land cover patterns

1. List some of the obvious land cover transformations.
2. What might the hills and slopes have looked like before being used for rice cultivation?
3. Explain why the slopes are terraced.
4. How do you think water is distributed between these terraced paddy fields?
5. What do you think are the main interruptions to some of the Earth's physical systems? Explain.
6. Would you consider these areas 'degraded'? Explain.

FIGURE 1.20 Longji rice terraces, Guangxi province, China

Activity 1.5b: Impacts of affluence and technology on land transformation

Refer to table 1.3 to answer the following questions.

TABLE 1.3 Causes and effects of common land cover transformations/interruptions

Causes of land and water degradation	Effects on Earth's physical systems	Example
Water removal and irrigation	Salinity, algal blooms and impact on aquatic creatures	Algal blooms and fish deaths in the Darling River, NSW; salinity problems in lower Murray catchment
Dam construction	Interruption to natural flow of river, effects on riparian areas downstream, loss of species	Three Gorges Dam, Yangtze River, China
Deforestation for grazing, cultivation, logging, roads and urban development	Loss of wildlife habitats, soil erosion, soil acidification, raising of water table, reduction in transpiration of moisture into air, reduction in O_2–CO_2 exchange	Amazon region in Brazil (cattle grazing and logging) and Sumatra, Indonesia (palm oil)
Overgrazing and overcultivation	Loss of ground cover, soil erosion, desertification	Sahel region, south of Sahara Desert, Africa
Chemical pollution from discharge and collapse of tailings dams	Water pollution, soil toxicity, loss of wildlife	Polyfluoroalkyl in Huron River, Detroit, US and lead in air at Mt Isa, Queensland
Waste disposal from landfill or illegal dumping, including litter	Water pollution, visual pollution	Plastics in Pacific Ocean

Explain and interpret information

1. Read through the types of land transformations in table 1.3. Write an extended paragraph on each one explaining how each is connected to 'growing affluence' or 'advances in technology'.
2. What strategies have been introduced in some countries to reduce the impacts of these transformations on people and the environment?
3. In which category would you place the examples in figure 1.21?

FIGURE 1.21 (a) A bean farm in south-east Queensland, (b) a cattle ranch, (c) dam construction in a river valley and (d) a rock quarry

Source: Bill Dodd

You can investigate graphs on how people have used the Earth's surface throughout time by using the weblink **Land cover over time** in the Resources tab.

Resources

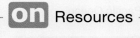 **Weblink** Land cover over time

Activity 1.5c: Comparing remotely sensed images to identify land cover change
Explain and analyse data

FIGURE 1.22 Changes to the Aral Sea, 1992–2015

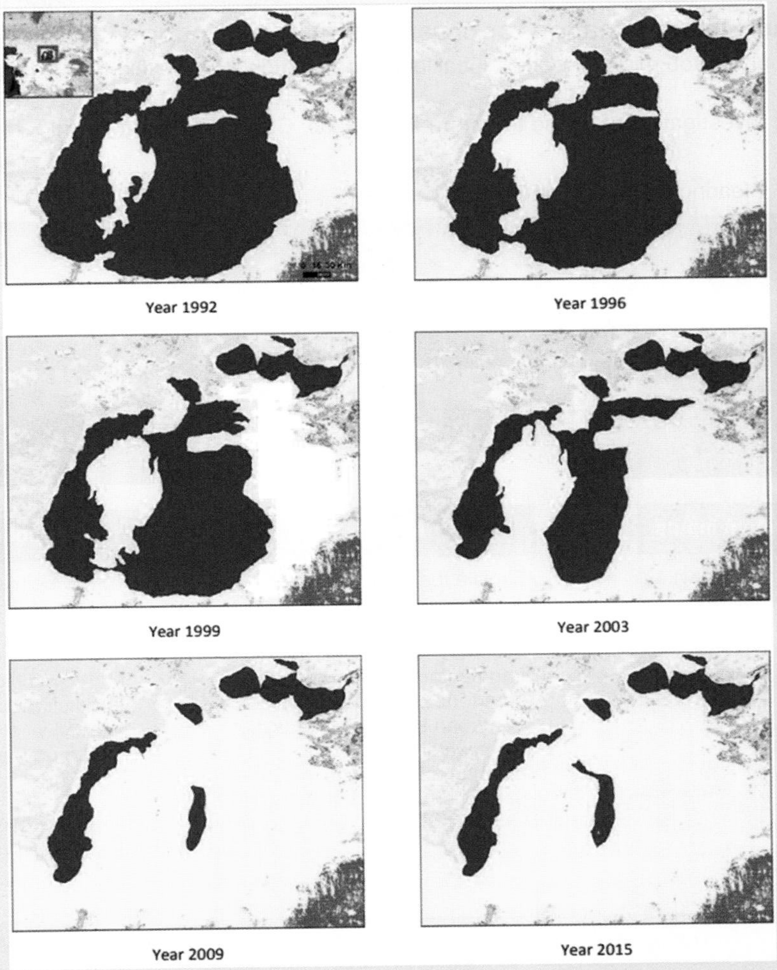

Year 1992

Year 1996

Year 1999

Year 2003

Year 2009

Year 2015

Source: © ESA Climate Change Initiative - Land Cover project led by UCLouvain 2017

Located between Kazakhstan (north) and part of Uzbekistan (south), the Aral Sea was once the fourth largest inland water body (lake) in the world. Covering an area close to 68 000 km^2, this vast inland sea was fed by two rivers – the Amu Darya and Syr Darya. The lake was once rich with fish, and adjacent shores supported thousands of desert nomads. Since the 1950s, diversion of river waters for irrigation and farming has had both a positive and negative effect. Although population and food production increased in the short term, this transformation resulted in the shrinking of the sea to only a fraction of its original area. Heavy use of fertilisers and pesticides caused water to become polluted, resulting in a collapse of fish populations. Soils were also affected by salinity and the exposed sea floor is a source of dust plumes and storms. Perhaps world heritage listing might be the only way to save the sea?

1. Study the images in figure 1.22. Make a list of all the changes you can see that may have occurred between 1992 and 2015.
2. Between which years do you think the most water was lost? Discuss with others in your class.
3. How have these changes interrupted the world's physical systems (e.g. hydrosphere — lake water has been extracted for irrigation or dried up from drought)?

> **Resources**
>
> **Video eLesson** Skillbuilder: Comparing aerial photographs to investigate spatial change over time (eles-2750)
> **Interactivity** Skillbuilder: Comparing aerial photographs to investigate spatial change over time (int-3368)

Activity 1.5d: Land transformation for urbanisation and transport
This is an example of how you might choose a local land modification issue, examine it and communicate it to the class.
  Local field study: Investigation of a land transformation issue: Clearing a forest corridor for a highway

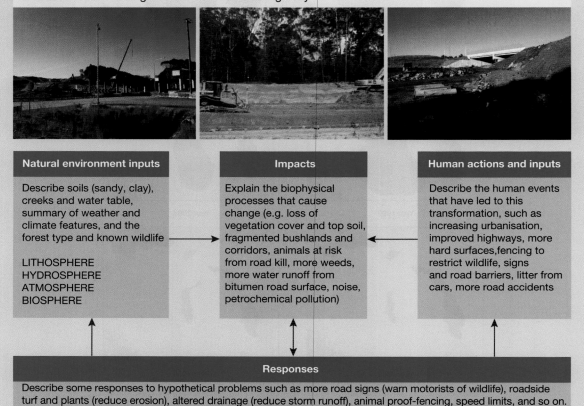

FIGURE 1.23 Clearing a forest corridor for a highway

Natural environment inputs
Describe soils (sandy, clay), creeks and water table, summary of weather and climate features, and the forest type and known wildlife

LITHOSPHERE
HYDROSPHERE
ATMOSPHERE
BIOSPHERE

Impacts
Explain the biophysical processes that cause change (e.g. loss of vegetation cover and top soil, fragmented bushlands and corridors, animals at risk from road kill, more weeds, more water runoff from bitumen road surface, noise, petrochemical pollution)

Human actions and inputs
Describe the human events that have led to this transformation, such as increasing urbanisation, improved highways, more hard surfaces, fencing to restrict wildlife, signs and road barriers, litter from cars, more road accidents

Responses
Describe some responses to hypothetical problems such as more road signs (warn motorists of wildlife), roadside turf and plants (reduce erosion), altered drainage (reduce storm runoff), animal proof-fencing, speed limits, and so on.

Source: Bill Dodd

1. Choose a local example where human activity has put pressure on part of the Earth's systems. Examine details of various interactions between natural and human systems, and record/photograph the levels of impact in terms of environmental, social and economic criteria.
2. Use your evidence to show if human responses to mitigate or restore environmental quality were successful.

1.5.3 Land cover changes in Australia since European settlement

European settlers have had a significant impact on Australia's land cover. When European colonisers arrived in Australia in the late 18th century, there was an urgent need for the colony to be economically self-sustaining to ensure survival. Land was cleared for farms, towns and roads, and graziers were encouraged to take herds of sheep and cattle further inland, searching for native pastures. However, as they built more dams and waterholes, they inadvertently helped kangaroo and rabbit populations increase, and the spread of livestock led to a rapid depletion of ground cover, and increased soil powdering and erosion (see figure 1.24). Occasional 'good seasons' also encouraged some graziers to overstock. When food for the livestock became scarce, some landowners either tore down trees or lopped mulga branches so that stock could feed on the leaves. Others ringbarked trees to thin out plant communities and encourage grass growth.

Large-scale scrub clearing also occurred in central Queensland's brigalow belt. Brigalow, a form of native wattle (*Acacia harpophylla*), grows in the drier margins (500–800 mm rainfall) between the Darling Downs and the Bowen Basin. Brigalow was regarded by Europeans as useless scrub with little economic value, so large pockets were cleared for grain growing and pasture in the Dawson and Callide river valleys. Today, less than one per cent of the original six million hectares remains in **national parks** or as remnant patches in state forests, private property or alongside roadways. Biologists now realise that brigalow and mulga ecosystems support many birds, small marsupials and reptiles, as well as hold soil together.

FIGURE 1.24 The impact of land clearing on the dry country of Central Queensland

Activity 1.5e: Causes of land cover change in Australia

1. Explain how European farming practices and changes to land use in Australia interrupted the interconnections between Earth's physical systems.
2. Make generalisations about the lasting impacts of these interruptions to Australia's current land cover.

1.6 Anthropogenic activity and how it has transformed land cover

1.6.1 Anthropogenic biomes

As the Earth's population grows, and more people need shelter, food, water and other resources, the number of unexplored or uninhabited places remaining in the world lessens. To satisfy these human demands, people are transforming much of the terrestrial surface. The global population is continuing to grow, but by how much and how quickly? How long can the Earth sustain population growth? How many people can the Earth support? What happens if there are too many people?

Studies have revealed that more than three-quarters of the Earth's ground surface has now been directly affected by human activities and that 24 per cent of the Earth's surface area is most likely to experience a decline in ecosystem function and productivity. At the same time, ocean and atmospheric studies show that the effects of pollutants are universal.

Areas that have experienced sustained human interaction are called anthropogenic biomes.

The physical spread of people around the world has been studied multiple times. German botanist-climatologist Wladimir Köppen mapped the original climatic zones in the early 1900s. These maps showed

FIGURE 1.25 Some effects caused by population growth and demand for space and resources

significant sections of the land surface to be unoccupied or pristine wilderness. However, a study by Erle Ellis and Navin Ramankutty in 2008 revealed that up to 77 per cent of the terrestrial surface is now human-dominated. They have re-mapped these human-altered landscapes into a mosaic of anthropogenic biomes, and their research has also shown that with technological advantage, people can live almost anywhere, apart from the most extreme environments. The only remaining wilderness areas are in very isolated mountains, forests, hot deserts or icecaps.

People use land resources for a range of activities, including agriculture, urban development, manufacturing, transport, water supply, forestry, coastal and port functions, energy and mining. Using three criteria — **population density**, land use type and common vegetation — Ellis and Ramankutty identified six major anthropogenic biomes. These are:

- dense settlements
- villages
- croplands
- **rangelands**
- forests
- wildlands.

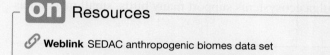

Resources

Weblink SEDAC anthropogenic biomes data set

1.6.2 Land cover change and climate change — is there a connection?

The effects of landscape change have become more noticeable as populations have increased and people have moved into what were once considered isolated or inhospitable environments. Over the past 200 years, the Earth's cover has been subjected to a great deal of alteration. Most changes have occurred in the land cover biomes. These changes include:

- forest and scrubland clearing for farming cultivation, cattle grazing, plantation **crops**, industrial development and urban expansion
- overgrazing and vegetation removal in arid areas resulting in desertification
- dam construction and diversion of waterways
- pollution of the atmosphere and waterways
- loss of wildlife habitats and depletion, and extinction of many wildlife species
- storage and careless dumping of solid and liquid toxic wastes
- extraction of natural resources such as **minerals**, timber and marine creatures.

Activities such as deforestation, desertification, land drainage, land reclamation, resource extraction, intensification of agriculture and pastoralism, coastal modification, and soil and water degradation can interrupt natural processes and systems. In recent times, land clearing and deforestation have been linked to desertification and an increase in surface albedo (see section 1.4.7). This results in a cooling effect, particularly in mid–high latitudes, and eventually lower rainfall. Deforestation also lowers **evapotranspiration**, again contributing to lower rainfall because less moisture is released into the air. These events may be contributing to global warming, climate change, extreme weather events, sea level rise, glacial and ice cap melting and coral reef deterioration.

Forests are very important in regulating climate change. Like oceans, forests act as a carbon sink that absorb carbon dioxide and other greenhouse gases that would otherwise remain free in the atmosphere. When large areas of forest are destroyed, this vital role as a sink is lost. It is estimated that approximately 15 per cent of all greenhouse gas emissions are the result of deforestation. As well, trees protect the topsoil with their roots and overhead canopy by reducing the impact of heavy rain. When trees are removed, so too is the protection for soils, which are easily weathered and washed into rivers. As nutrient-rich soils are washed away, people think they need to clear more land to grow crops. Land activities such as farming requires ploughing of soil, grazing and timber extraction, and these result in soil loss, and mining and quarrying require digging and movement of soil. All of these contribute to soil erosion. It is easy to forget that this same soil is essential for future generations.

FIGURE 1.26 Prime farming land can be destroyed by mining.

Australia's diverse climate zones, topography and soils support a wide range of land covers and uses. However, land cover change in Australia is different from many other industrialised countries where land usage patterns have become relatively stable. In Australia, land use patterns are still undergoing significant change, with approvals still being given to the farming and forestry sectors to 'open up' more land and retrieve timber. At a time when coal-fired power stations are being phased out in many countries, governments in Australia are still considering opening huge coal mines.

1.6.3 Forests — the dominant land cover

Forests are the largest and most widespread of all biomes. Dominated by trees, forests cover approximately 31 per cent of the land surface, support the most terrestrial **biodiversity** and contain up to 80 per cent of the total plant biomass.

FIGURE 1.27 Location of the world's forests

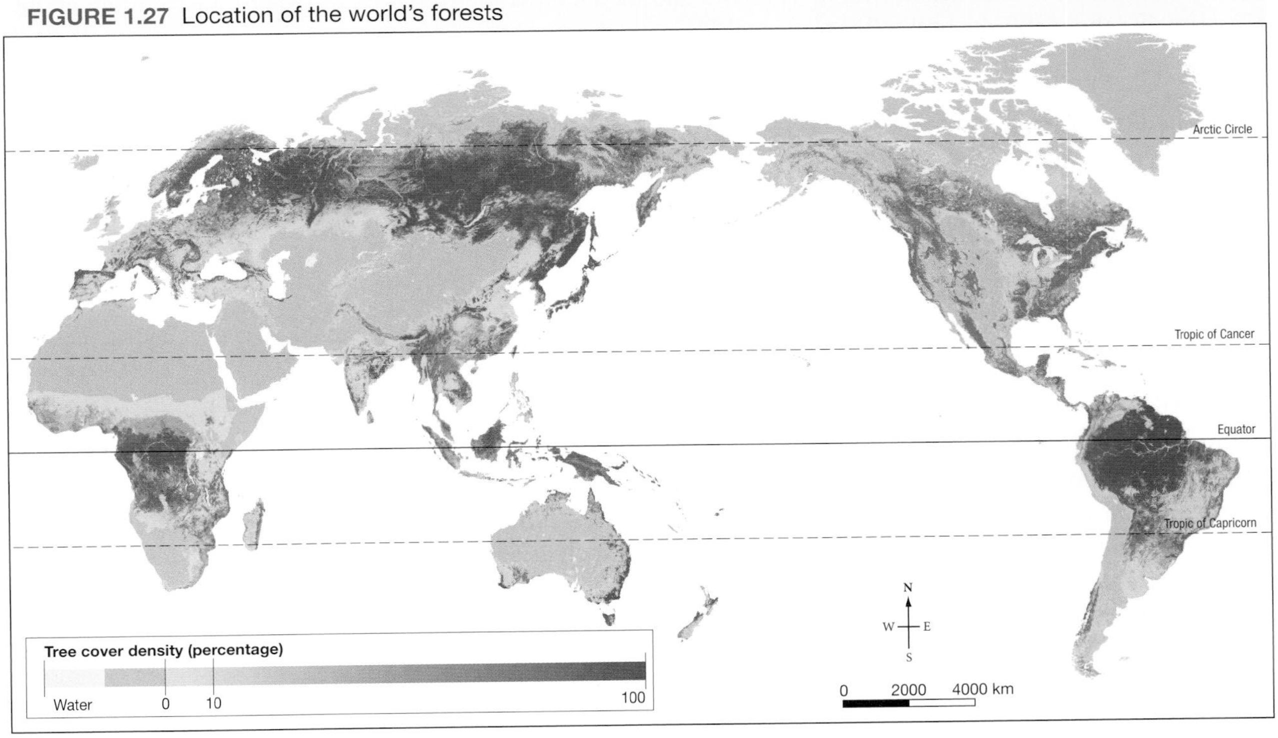

Source: Food and Agriculture Organization of the United Nations, 2010, FAO Data, http://foris.fao.org/static/data/fra2010/forest2010mapwithleg.jpg. Reproduced with permission.

Forests are the most complex terrestrial biome. Depending on the climate, soil, aspect and elevation, they support more than 60 000 different species of trees. They also provide timber to people for building and may be used as national parks.

The trees in forests have many uses. It is estimated that approximately 1.6 billion people rely on forests for food, fresh water, timber, clothing and traditional medicine. However, a forest's greatest asset is probably the ability for **carbon sequestration**. A healthy growing tree can absorb about 20 kg of CO_2 each year. A tree that has lived for 40 years will have stored at least a tonne of CO_2.

FIGURE 1.28 Eucalyptus forest

Source: Bill Dodd

Forests can be classified in many ways. There are tropical and sub-tropical rainforests in the warmest regions, temperate forests in the mid-latitudes, and boreal coniferous forests in the colder climates (refer back to figure 1.19). Geographers group them by biome (e.g. rainforest or open forest) and biogeographic area (e.g. mangroves or mulga). Others, such as botanists, may group them according to leaf type (e.g. evergreen, deciduous or coniferous).

Anthropogenic activity is reducing the amount of forest land cover all over the world. A hundred years ago, the world supported about 50 million km^2 of forest, but today that has declined to about 40 million km^2. Evidence also reveals that deforestation is occurring at an alarming rate. According to the **Food and Agricultural Organization (FAO)**, approximately 7.3 million hectares are destroyed each year. These are mostly rainforests in Indonesia, Brazil, Thailand and the Democratic Republic of Congo. Learn more about forests using the weblinks in the Resources tab.

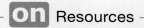 Resources

🔗 **Weblinks**   Percentage of forest as land area
World map of the earth's vegetation
Global forest change map
The State of the World's Forests 2018

Rainforests

Rainforests are thriving communities of plants that have adapted to very high levels of rainfall and humidity, and have attained a state of **ecological climax**. They grow in hot, equatorial countries including Brazil, Indonesia, India, Malaysia, Papua and New Guinea, Zaire and northern Australia (see figure 1.29). Temperate rainforests also flourish in wet, cooler places such as southern Queensland and New South Wales, and colder areas like Tasmania or New Zealand. Rainforests cover about 6 per cent of world's land surface and are believed to produce up to 40 per cent of the Earth's oxygen.

FIGURE 1.29 Introduction to tropical rainforests

Tropical rainforests are found in the warmest and wettest countries of the world such as Brazil, Indonesia, India, Malaysia, Papua and New Guinea, Zaire, Peru, Gabon and northern Autralia. Rainfall totals may vary from as little as 250 mm / year to as mush as > 3000 mm / year. Rainfall is generally constant while humidity remains high. Temperatures rarely drop below 20°C.

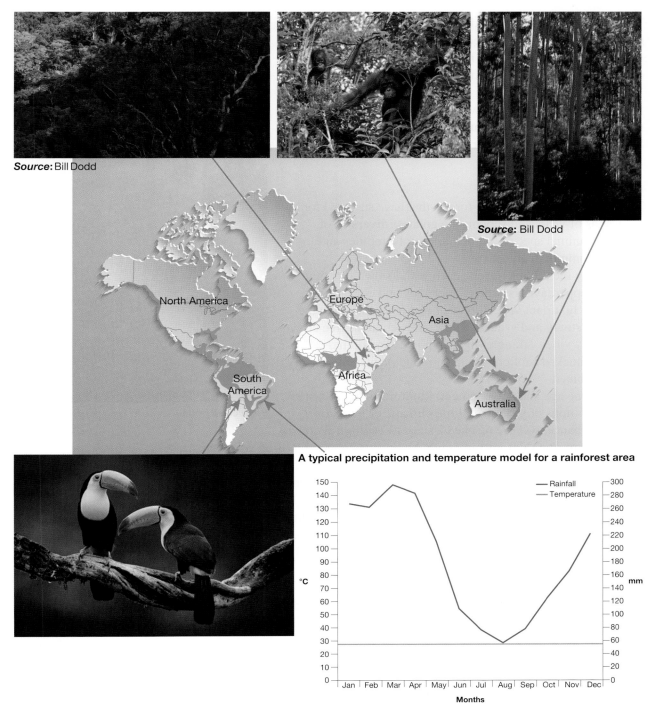

Source: Bill Dodd

Source: Bill Dodd

Source: © Climate-Data.org / AM OP / OpenStreetMap contributors

The term 'tropical rainforest' was first used by the German geographer A. F. Schimper in 1898. He described *Tropische Regenwald* (tropical rainforest) as 'evergreen, hygrophilous in character, at least thirty metres high, but usually much taller, rich in stemmed lianes and in woody as well as herbaceous epiphytes.'

Location

Tropical rainforests are found in regions adjacent to the equator due to the interconnections between global climatic systems (see subtopic 1.4). The main clusters are located in central and northern South America, south-east Asia, central Africa, and Australia and the Pacific islands.

Temperate rainforests grow in cooler coastal strips along the Pacific coast of North America, coastal areas of Chile in South America, and in cooler parts of southern Australia and New Zealand.

Appearance and features

Tropical forests are primeval forests that have survived for thousands of years and support the greatest **diversity** of plant and animal species, particularly birds, reptiles and insects. Common features are:

- clearly identifiable layers, each with its own function. The upper layer is called the canopy and consists of dense luxuriant vegetation that restricts sunlight to the understorey and ground. The canopy contains tree crowns, woody vines, strangler figs, epiphytes and orchids. The blocked sunlight ensures ground plants are sparsely distributed.
- extreme age and development. Rainforests have existed for millions of years with little climatic or human disturbance, so have reached a stage of ecological climax. The Amazon rainforest in South America is about 55 million years old.
- a highly diverse variety of plant and animal species. Rainforests support over half of all known animal species and over two-thirds of known plant species. On average, each square kilometre supports as many as a hundred different tree species.
- a self-sustaining decomposing layer of leaf and twig litter that maintains nutrient recycling. Each day, tonnes of leaves and twigs fall from the canopy and decay within two to five months, adding minerals to the topsoil. Decomposition is assisted by fungal and bacterial species that thrive in the warm wet conditions. Breakdown at the ground layer is ongoing, as much of the added nutrient is leached away by heavy rainfall, making the soils slightly acidic.
- unique plants that have adapted to the wet, humid conditions, such as the massive buttress roots needed to support huge trees (up to 35 m) in damp shallow soils to prevent them being toppled by extreme winds, and leafless woody stems known as cauliflory growing out of tree trunks.
- humid, shady conditions that create a microclimate.

FIGURE 1.30 Rainforests feature a closed, shady and damp environment and luxuriant vegetation.

Source: Bill Dodd

Weather and climate

Rainforests are characterised by hot/warm and wet weather most of the year. Temperatures seldom fall below 18 °C and humidity is generally high, often greater than 75 per cent. There are two distinct seasons: wet and dry. In monsoonal areas, the dry period may be longer, but there is about 12 hours of daylight, regardless of the season. Rainfall is between 1500 mm and 5000 mm, often as heavy thunderstorms.

Vegetation

Rainforests feature very large, tall trees that provide food and many other resources. Due to the wet soil, the trees need buttress roots to support them. They also provide a canopy that blocks up to 95 per cent of the sunlight from the forest floor, and an understorey that contains palms, ferns, lianas, strangler fig, epiphytes and orchids. Approximately 3000 edible fruits grow in rainforests, many with medicinal capabilities. Several trees provide everyday goods, such as coffee, rubber, bananas, mangoes and figs, while others, including the cedar, maple, teak and mahogany varieties, provide valuable timber.

Wildlife

Due to their wide range of locations, rainforests support many different species of mammals, birds, insects, reptiles, amphibians, fish and worms. Some of these include the gorilla in Africa, the jaguar in South America and the orangutan in south-east Asia. It is estimated that half of the 10 million known species of animals, insects and plants live in rainforests. The majority of these are insects, such as butterflies, beetles and mosquitoes. Many of the other animals live in the trees due to predators, and some rarely descend to the forest floor.

Human activities

Rainforests are important environments for humans for various reasons. Most importantly, rainforests help mitigate carbon emissions and regulate the O_2–CO_2 balance — scientists estimate that rainforests can absorb 210 gigatons of carbon. Some countries have indigenous communities that live in the rainforest, and these people see the forest as their home, which provides shelter and food. In other countries, remaining rainforests are protected as national parks or wildlife reserves, which people can visit as a form of recreation. Many scientists study rainforests because they are places of medical and biological learning. However, many rainforests are destroyed by human activities, such as either legal or illegal logging.

1.6.4 Using forests for recreation and national parks

National parks are areas of significant worth, unique biodiversity or unusual landforms. They are protected by law and have special conditions for people who visit or access them. They are often part of community environmental management projects, which involve setting aside areas for the public to enjoy and preserve for future generations.

Rainforests are now rare in many countries, so it is common for them to be recognised as national parks or heritage sites. To ensure they are not damaged and their resident wildlife protected, people have limited access and visiting conditions. National parks are important places where people can learn and enjoy the experience of a forest in its original state and help the community appreciate the need to retain such ecosystems. As well, national parks and other forests may be preserved as environmental parks, scientific study areas, fauna sanctuaries, or be listed as part of the national estate.

Some well-known forest national parks in Queensland are Lamington, Eungella, Daintree and Springbrook. The Gondwana Rainforests of eastern Australia total 366 500 ha and attract up to 2 million visitors each year.

1.6.5 Clearing of rainforests

Extensive forest clearing over the past 200 years has led to significant transformation of many rainforest areas. Known as deforestation, this process refers to the intentional clearing of forests to use the land for other purposes, such as farming, cattle grazing, logging, mining, urban and industrial development, as well as the construction of dams, motorways and airports. Once cleared, these forests are unlikely to ever be restored. As a result, an area transformed from a carbon sink to a non-carbon purpose alters the carbon cycle (see section 1.3.1), creating an imbalance of greenhouses gases in the atmosphere, which ultimately leads to an enhanced greenhouse effect (see section 1.4.8).

Deforestation occurs in a variety ways. These include burning, clear-felling for agriculture, ranching and development, logging for timber, and degradation due to climate change. When a forest is deliberately torched to make it easier for clearing, deterioration occurs quickly. Shade tolerant plants are unable to grow, soil fertility from leaves is lost, wildlife is killed or dies from starvation or loss of protective habitat, soil erosion is highly likely and the forest's sequestration properties are gone. Additionally, there are fewer trees to transpire moisture back into the atmosphere, which would have adverse impacts on global precipitation patterns.

Forest clearing is currently occurring on a large scale in places like Brazil and Indonesia. Trees are cleared and replaced with livestock ranching and farming, plantation agriculture (palm oil and coffee) and mining. According to the FAO, almost 8 million ha of forest are cleared each year. According to the World Wide Fund for Nature (WWF), this equates to approximately 27 football fields per minute. In 2016, fires were particularly destructive in Brazil and Indonesia, destroying a combined area about the size of New Zealand. Even though the fires happened during dry periods, most were deliberately lit as part of land clearing programs. While deforestation has recently decreased in some countries, any gains have been offset by increased numbers of large wildfires due to extreme weather and possible climate change.

FIGURE 1.31 The effects of deforestation

Healthy ecosystem
- Trees intercept rainfall to protect forest floor
- Leaves fall to ground to supply nutrients to soil
- Bacteria and fungi convert litter to humus
- Sunlight activates photosynthesis and food taken in by roots
- Leaves absorb CO_2 and release oxygen into the air
- Carbon sequestered by wood
- Very little soil erosion
- Habitat for many animals and birds

After deforestation
- No trees to reduce impact of rainfall; significant soil erosion
- No leaves to nourish soil
- No humus replacement
- Carbon released when forest burnt; no oxygen produced
- Increased loss of nutrients by leaching
- No longer habitat for wildlife
- No evapotranspiration and therefore less moisture for precipitation

Deforestation often occurs to make way for animal grazing, agriculture and mining (Brazil), plantations for palm oil (Indonesia and Borneo) and mining (Brazil). For example, in Brazil, pasture land has more value than forested land, so clearing is the preferred option for both farmers and land speculators. Approximately 70 per cent of forest clearing in Brazil is for cattle grazing. Brazil also clears forested areas so mining can occur. Deforestation is widespread in Indonesia to allow companies to grow palm oil for biofuel and food additives. Nearly 40 per cent of the world's certified palm oil comes from these areas.

Read more about palm oil production using the weblink in the Resources tab.

 Resources

 Weblink Palm oil

Activity 1.6a: Explaining the motives and effects of large scale deforestation

Explain, synthesise and analyse information

1. Referring to figure 1.31, compare the differences between a healthy rainforest ecosystem and one where trees have been removed and the ecosystem eliminated.
2. Write two extended paragraphs to explain these differences by using the key points listed in your sentences.
3. In a third paragraph, attempt to generalise how these changes might impact on the future of a local anthropogenic biomes, such as a forest park.
4. Choose **ONE** of these reasons for deforestation mentioned in the information box about deforestation in Brazil, Indonesia and Borneo, and investigate it further. Complete a comparison table to show advantages and disadvantages by inserting recent data or events associated with the activity.

Criteria	Advantages	Disadvantages
Economic (local and national)		
Social (local communities)		
Environmental (plants, soils, wildlife)		

5. Find out why such large sections of forest are cleared for these activities. Refer to the websites in the Resources tab for more information.

Resources

Weblinks Mining
Palm oil plantations
Grazing and agriculture

Activity 1.6b: Ethics of protecting pristine areas

Communicate, explain and analyse information

In May 2018, a study was released that showed that one-third of world's protected areas have been degraded by human activities.

1. Refer to a summary of this report by going to the weblink **World's areas degraded** in the Resources tab.
2. In a paragraph, summarise the intent of the report by quoting key words or phrases.
3. Explain the terms *not fit for purpose* and *intense human pressure*.
4. What activities are causing these pressures?
5. How serious is the problem?
6. Which countries are the worst offenders, according to the report? Examine the global footprint maps at the **Human footprint** weblink in the Resources tab and comment on where human pressure appears to be greatest.
7. What do you think should to be done to save protected areas?

Resources

Weblinks World's areas degraded
Human footprint

Activity 1.6c: Spatial patterns of land cover
Comprehend and analyse data to identify trends and patterns

FIGURE 1.32 Tropical deforestation by location of the world's rainforests

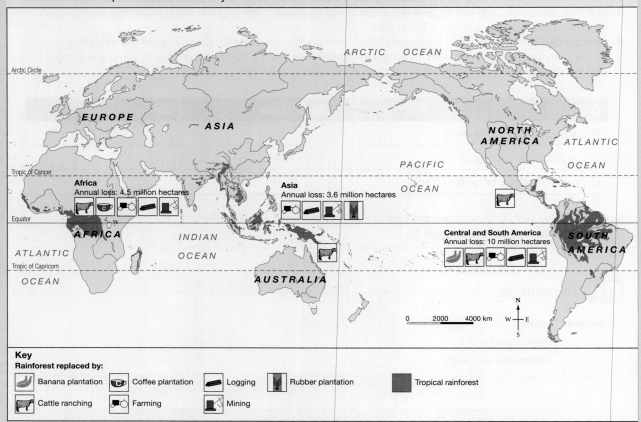

Source: MAPgraphics Pty Ltd, Brisbane

FIGURE 1.33 Change in forest cover over time: (a) 2001 (b) 2018

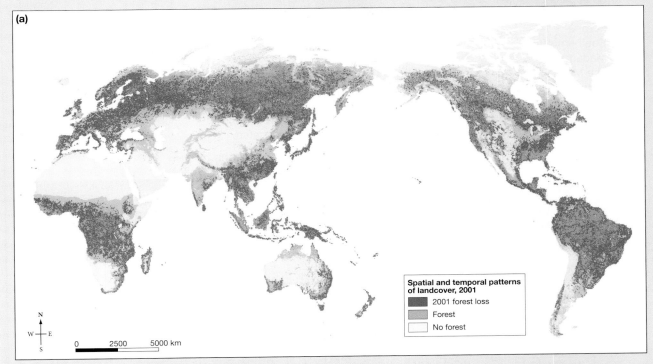

Source: Hansen / UMD / Google / USGS / NASA

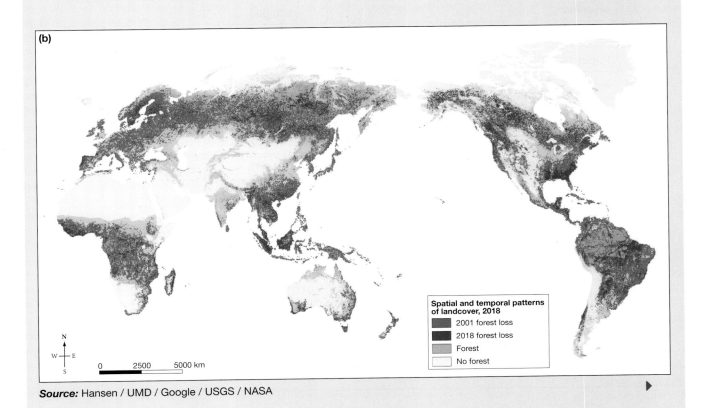

Source: Hansen / UMD / Google / USGS / NASA

1. Refer to changes to forest cover as shown in figures 1.32 and 1.33. Identify and name, if possible:
   (a) areas that have changed considerably
   (b) areas that have changed a little, but not considerably
   (c) no apparent change.
2. Look closely at the forest cover in the following places and comment on any changes.
   (a) Europe
   (b) Southern Africa
   (c) Madagascar
   (d) The Middle East.
3. What do you think caused the most forest depletion in north-east North America and eastern Australia?
4. Examine the list of countries with large areas of rainforest in table 1.4. Complete the table by researching current areas of rainforest and the measure of loss or gain.
5. Using an atlas, identify on a world map where these countries are located.

TABLE 1.4 Levels of rainforest destruction

#	Country	Continent or region	Rainforest area 200 years ago (km^2)	Current rainforest area	Loss or gain area
1	Brazil	South America	2 860 000		
2	Congo, Dem Rep	Africa	100 000		
3	Indonesia	South-east Asia	1 220 000		
4	Colombia		700 000		
5	Peru		700 000		
6	Venezuela		420 000		
7	Myanmar		500 000		
8	Bolivia		90 000		
9	Papua New Guinea	Australian Pacific	425 000		
10	India		1 600 000		
11	Mexico		400 000		
12	Suriname		125 000		
13	Guyana		120 000		
14	Madagascar		62 000		
15	French Guiana		120 000		
16	Congo		100 000		
17	Ecuador		132 000		
18	Thailand		435 000		
19	Malaysia		305 000		
20	Zaire		1 245 000		

Activity 1.6d: Processes affecting forest land cover

Interpret models and graphs

1. Examine the general nutrient recycling model shown in figure 1.34a.
2. Using the terms provided and considering the size of the circles, explain how nutrients are recycled in a rainforest using figure 1.34b.
3. Justify why nutrient recycling in a boreal forest (pine trees) is much lower than in a rainforest by comparing it to figure 1.34c.
4. The most common causes of deforestation in the Amazon rainforest are cattle ranching (about 65 per cent), small- and large- scale agriculture (about 30 per cent), legal and illegal logging (3 per cent) and Other (2 per cent).
   Using this data, construct a pie graph to compare levels of these activities as in figure 1.35.
5. The data in table 1.5 shows the extent of deforestation since 2000.
   Which five years were most destructive? What was significant about 2009 levels? Try to find out why.
6. What is the trend with the total area of forest lost since 1970? Try to find out what events or policy changes occurred in Brazil around 1970 to makes this a benchmark year.
7. Use the data in column 2 of table 1.5, draw a line graph showing the trend of the decreasing remaining area of rainforest.

FIGURE 1.34 Nutrient cycling: (a) general model (b) tropical rainforest (c) boreal forest

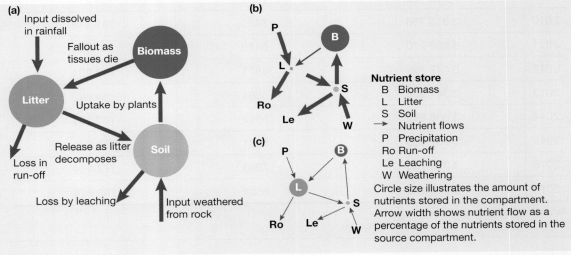

FIGURE 1.35 Causes of deforestation in the Amazon

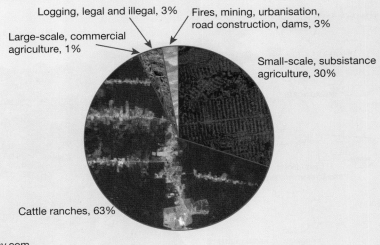

Source: © mongabay.com

TABLE 1.5 Extent of deforestation since 2000

Year	Estimated area of rainforest in Amazon (km^2)	Annual deforestation (km^2)	Total area of forest removed since 1970
2000	3 524 097	18 260	575 903
2001	3 505 932	18 165	594 068
2002	3 484 538	21 651	615 719
2003	3 459 291	25 396	641 115
2004	3 431 868	27 772	668 887
2005	3 413 022	19 014	687 901
2006	3 398 913	14 285	702 186
2007	3 387 381	11 651	713 837
2008	3 375 413	12 911	726 748
2009	3 365 788	7464	734 212
2010	3 358 788	7000	741 212
2011	3 352 370	6418	747 630
2012	3 347 799	4571	752 201
2013	3 341 908	5891	758 092
2014	3 336 896	5012	763 104
2015	3 331 065	5831	768 935
2016	3 322 796	7893	777 204
2017	3 316 172	6624	783 828

Source: © mongabay.com

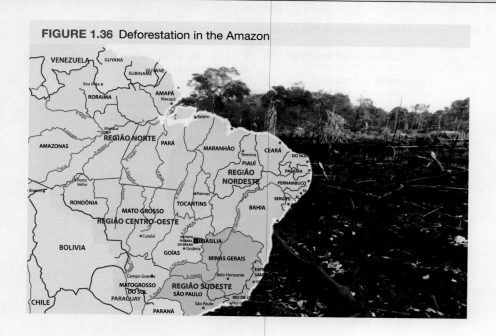

FIGURE 1.36 Deforestation in the Amazon

Resources

- **Video eLesson** Skillbuilder: Constructing a pie graph (eles-1632)
- **Interactivity** Skillbuilder: Constructing a pie graph (int-3128)
- **Weblink** 30 years of deforestation in Amazon rainforest

1.6.6 Tropical grasslands (savanna) and rangelands

The term **savanna** refers to a biome classification and was first used by the people in Central America when referring to flat treeless areas where they lived. Today, it is used to describe similar grassy and scattered tree plains anywhere in the world, but particularly the grasslands of eastern Africa and northern Australia. Many savanna areas have been converted into rangelands, a general term that also includes shrublands, woodlands, and semi-desert areas used for grazing domestic livestock or wild animals. Rangelands is also used to describe both tallgrass and shortgrass **prairies**, **steppes**, **chaparrals**, and some tundras if used for animal grazing.

Savanna lands form a transition zone between forest regions and hot deserts. Because biomes are controlled by climate, they don't have definite boundaries. Instead they gradually merge into different landscapes, sometimes having features of both biomes. Savanna regions have some trees but also dry grassy features.

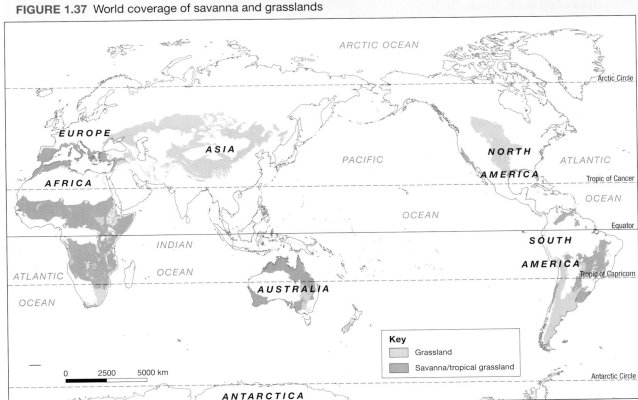

FIGURE 1.37 World coverage of savanna and grasslands

Source: MAPgraphics Pty Ltd, Brisbane

Location

As figure 1.37 shows, savanna areas and grasslands are found all over the world, but they cover the largest areas in Africa, central Asia, South America and Australia. They border desert regions and are most commonly found between the tropics of Cancer and Capricorn where there are areas of high pressure (see section 1.4.1).

Appearance and features

Savanna and grassland areas are extensive plains, covered by tall, coarse grasses that stand up to 4 m high. There are occasionally hills and mountains, and intermittent stands of deciduous trees. Some of these trees include the Umbrella Thorn Acacia, a hardy tree that can endure extreme temperatures, and the Boab tree, with its peculiar thick trunk.

Weather and climate

The climate in savanna areas and grasslands tends to be tropical, with average monthly temperatures between 20 and 32 °C, and high daytime ranges. The summers are very wet, with heavy rainfall; winters are dry and drought-like. The annual rainfall is be between 1000 and 1500 mm.

Vegetation

The vegetation in savanna and grassland areas is typically coarse and low to the ground. There are tall, tropical grasses that are drought-tolerant and fire resistant, and thorny plants and shrubs. Deciduous trees appear sporadically and many have adapted to store water, such as the Boab tree found in northern Australia and parts of Africa. During the dry season, many plants wither and die, then regrow when the rain comes.

FIGURE 1.38 Savanna areas contain grasses and intermittent vegetation.

Wildlife

Savanna areas and grasslands support a diverse range of animals, both carnivorous and herbivorous. In Africa, there are herds of grazing animals including giraffe, elephants, zebra, wildebeest and buffalo. These animals are hunted by predators such as lions, cheetahs, hyenas and jackals. There are also river dwellers such as hippopotamus and crocodiles. As the dry season sets in, many animals migrate to find food and return to the grassy pastures when the rain comes.

FIGURE 1.39 African grasslands support a selection of large mammals.

Human activities

Human activities in savanna and grasslands vary from country to country. Many communities use these areas for **nomadic herding** and grazing of livestock, and parts of South Africa, Botswana and Namibia harvest wood, fruit and seeds. In Kenya, Tanzania and Uganda, savanna and grasslands are used as national parks and large-scale safari tourism, which brings money into the country. Mismanagement of anthropogenic activities in savanna areas, such as over grazing, make them prone to land degradation.

FIGURE 1.40 (a) Farmers and (b) agro-pastoralists in Zimbabwe

Activity 1.6e: African development proposal
Explain and analyse a proposal

1. Read the case study that follows and use the information from section 1.6.6 to complete a SWOT analysis of the proposal to transform savanna grasslands into agricultural cropland and rangeland grazing. A few sample ideas are included to help you get started.
   - Strengths
     Vast agricultural resource with considerable farming potential.
     Help reduce Africa's widening and increasingly expensive net food trade deficit.
   - Weaknesses
     Region has an important role in physical systems. For example, large expanse of grasses reduces flooding and erosion by slowing down surface run-off during heavy rains.
   - Opportunities
     This will help reduce food costs for many African people.
     Improved nutrition for mothers and babies.
   - Threats
     The project will put pressure on Africa's rare grassland wildlife with conflict over land use likely.
     Threaten gains made by safari tourism, which contributes $2.5 billion per year to Kenya's GDP and $404 million to Rwanda's GDP.

FIGURE 1.41 SWOT analysis

Strengths (+)	Weaknesses (−)
•	•
•	•
•	•
•	•

Opportunities (+)	Threats (−)
•	•
•	•
•	•
•	•

2. Read the following case study and consider the statement *The savanna grasslands are 'one of the major underutilised resources in Africa and should be exploited to help reduce hunger and provide more employment.'*
   Do you agree?
   Use the data from your SWOT analysis and debate the pros and cons in class.

PROPOSAL TO TRANSFORM SAVANNA INTO FARMLANDS

In October 2017, the African Development Bank raised a proposal to transform the African savanna into an agricultural and livestock grazing agribusiness hub. The initial goal is to cultivate about two million hectares of grasslands in eight different countries — Ghana, Guinea, the Democratic Republic of Congo, Central African Republic, Uganda, Kenya, Zambia and Mozambique.

The Bank's Vice-President for Agriculture, Human and Social Development, Jennifer Blanke, told an audience at a 2018 World Food Prize event in the US that the savanna regions were like a 'Sleeping Giant' that needed to awaken and become the continent's green revolution. She described the savanna as 'the world's largest agricultural frontier', which could reduce African dependency on overseas food imports and provide direct jobs for tens of millions of young people and indirect jobs for many more. Such a project would also help contribute to feeding the world.

Africa has about 60 per cent of the world's uncultivated arable land, but currently spends an estimated US$35 billion per year on importing food. Blanke said this vast area of cultivatable land could support the production of maize, soybean and livestock, and transform the continent into a net exporter of these commodities. She also claimed that Africa was importing foods it can produce domestically — 22 million metric tons of maize, two million metric tons of soybean, one million metric tons of broiler meat and 10 million metric tons of milk product each year.

Blanke claimed that transforming a small part of Africa's mixed woodland grasslands, in a smart and sustainable way, can produce enough food to supply all the continent's maize, soybean and livestock needs. She compared this development proposal with the success of Brazil, which developed its tropical Cerrados region to produce food. As well, African savanna soils are less acidic than those in Brazil and should be more productive.

Source: Adapted from African Development Bank Group

1.6.7 Deserts

The world contains large areas of hot, dry land known as desert. Deserts are areas that receive less than 250 mm of rain in a year and where land has less than 50 per cent of the ground surface covered by vegetation. It is estimated that deserts (hot and cold) cover approximately 20 per cent of the terrestrial surface. Hot deserts experience extreme sun and heat, due to their close proximity to the equator; cold deserts are generally found in mountainous areas between the polar regions and the tropics, which means they receive less sun and tend to be colder during winter. Despite their often harsh conditions, deserts are used by people for mining and extensive cattle grazing, scientific observations, remote weapons testing and military bases.

FIGURE 1.42 Erg deserts are dominated by large, sandy dunes that are shaped by the wind.

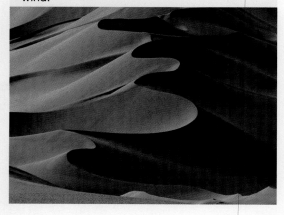

Location

Deserts are found on every continent except for Europe. Due to global climatic systems, hot deserts are always found between latitudes 15 and 30 degrees north or south of the equator (see subtopic 1.4), on the western side or middle of a continent, and in regions adjacent to large mountains where there is a **rain shadow**. There are several reasons for this, including:

- latitudes 15 to 30 degrees are dominated by high pressure cells. Because high pressure cells descend from the dry upper atmosphere, they prevent clouds forming so it cannot rain.
- the western sides of continents are adjacent to the world's cold ocean currents. Because the surrounding air is cold, it is unable to hold moisture or rise upwards to form clouds.
- high mountains tend to force moist onshore winds upwards, making them drop their rain. When they eventually pass they do not have sufficient moisture to form clouds on the **leeward side**, thus creating a rain shadow.

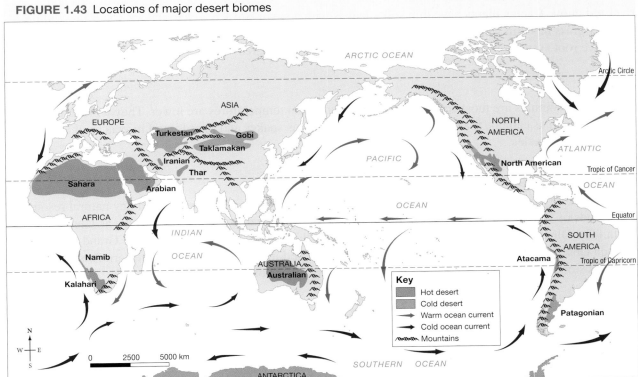

FIGURE 1.43 Locations of major desert biomes

Source: MAPgraphics Pty Ltd, Brisbane

Appearance and features

Deserts are very rocky and are therefore often classified according to their rock type. The most common desert types are Erg (sandy), Reg (stony) and Hamada (rocky). Some deserts contain large sale pans in low-lying areas where saline lakes have dried up (e.g. Lake Eyre, Lake Torrens and Lake Frome, all in South Australia). Many of Australia's deserts appear red or yellow, due to the hard crusts of iron (red) and aluminium oxides (yellow) that have formed on the surface.

Weather and climate

Deserts are characterised by extreme temperatures. Most of the year, deserts have hot days and cold nights, but in the summer, daytime temperatures can climb higher than 45 °C and night temperatures can drop below 0 °C due to the lack of cloud cover to retain warmth. There is very low rainfall most of the year (less than 250 mm annually) and this tends to come as storms so rapid runoff causes considerable erosion. Humidity is less than 20 per cent in most places, but may rise as high as 50 per cent after rainfall.

Vegetation

The plants found in deserts must be able to survive with minimal water, due to the very dry conditions. They are classified as either **drought escapers** or **drought resisters**.

Drought escapers lie dormant during dry periods. When rain falls, they germinate and grow quickly. They include saltbush and flowers known as ephemerals, whose bright and colourful blooms attract insects that ensure pollination and survival occurs.

Drought resistors are perennials (plants that grow all year) and have systems that help them retain water. These plants include cactus, spinifex, and mulga, and trees such as coolabahs, ghost gums and river red gums. Cactus plants have thin spiky leaves with a small surface area to reduce water loss, while spinifex grass has thin waxy leaves and grows in thick clumps to protect its root system. Mulga bushes have extensive root systems to retrieve moisture from a large surface area.

Wildlife

Many desert animals are most active during the night, when temperatures are lower. They include hardy mammals such as camels, foxes and bilbies, reptiles such as snakes and lizards, and insects such as scorpions and beetles.

Human activities

Deserts can form beautiful and unique landscapes ideals for tourism, and often hold many natural resources. Thousands of visitors travel to deserts all over the world every year to see landforms such as Uluru, Kata Tjuta (The Olgas), Wave Rock, Kings Canyon and Standley Chasm in Australia, and the Grand Canyon and Nevada Desert landforms in the US. Metals such as gold and iron ore are mined in Western Australia, and precious stones, including opals, are mined at Coober Pedy in South Australia.

Activity 1.6f: Uluru weather data

Explain and interpret data

Refer to figure 1.44 to answer the following questions.
1. What weather information is shown by the red line? Does it match with the information in table 1.6 (rounded off)?
2. What does the orange line show?
3. What information does the orange zone show?
4. What data is shown by the blue bars?
5. Estimate the mean monthly rainfalls and complete table 1.6.
6. What is the mean annual total in mm?
7. Does Uluru have a 'wet' season and 'dry' season?
8. Calculate the average annual total rainfall. Is Uluru desert or semi-desert?

FIGURE 1.44 Climate chart for one of the weather stations near Uluru

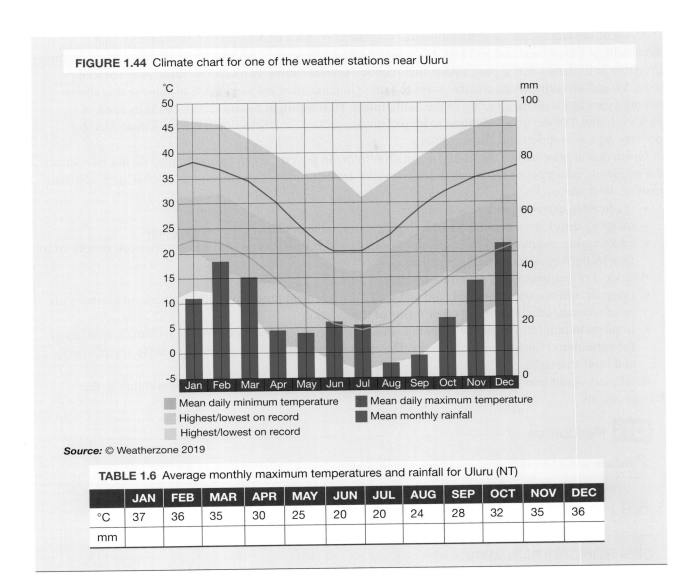

- Mean daily minimum temperature
- Mean daily maximum temperature
- Highest/lowest on record
- Mean monthly rainfall
- Highest/lowest on record

Source: © Weatherzone 2019

TABLE 1.6 Average monthly maximum temperatures and rainfall for Uluru (NT)

	JAN	FEB	MAR	APR	MAY	JUN	JUL	AUG	SEP	OCT	NOV	DEC
°C	37	36	35	30	25	20	20	24	28	32	35	36
mm												

1.6.8 Desertification

Desertification is a form of land degradation that transforms land that was once fertile and arable into unproductive arid land. In many places it is an expansion of an existing desert or dryland due to human activities (overgrazing or vegetation removal) or natural causes (drought). The continued expansion of desert areas through the process of desertification will have significant ramifications on albedo (see subtopic 1.4). Desert areas have some of the highest rates of albedo and any increase in size will subsequently increase the amount of heat being reflected back into the atmosphere. In conjunction with other anthropogenic activities that increase greenhouses gases in the atmosphere, such as the removal of vegetation, global temperatures will increase. This enhanced greenhouse effect will further exacerbate the rate of desertification due to subsequent changes to precipitation patterns.

FIGURE 1.45 Causes of desertification

- Soil is blown away by wind or washed away by storms
- Land is overgrazed with animals or over-cultivated
- Soil loses its structure and fertility
- Loss of vegetation and protective cover

Much of the large-scale desertification was originated by poor traditional farming methods that eventually led to erosion and soil loss, but reversing desertification is not just about stopping basic farming practices in drylands. It is a progression that requires several strategies including addressing societal attitudes and inventing more creative ways to farm. Genuine attempts were made to achieve this almost twenty years ago with the drafting of the **Millennium Development Goals** (2000–2015) by the UN. However, much of the good intention to help countries reduce desertification has been jeopardised by poverty, war, corruption and climate change.

Given that at least half of the global population living in poverty are in drylands, they do not have either the economic resources or technical skills to reverse desertification. Consequently, a global approach may involve these ideologies:

- eradicating extreme poverty and hunger
- assisting developing countries in developing sustainable land management programs
- encouraging international research, development and cooperation where all countries can benefit from new technologies and farming practices.

At a local or regional level, governments may:

- integrate soil conservation, and land and water management programs with strategies to protect soils from erosion, salinisation and other forms of degradation
- implement programs of desert greening, where strategies are developed for the reclamation of deserts for agriculture, forestry or biodiversity. Desert greening also has the potential to solve water, energy, and food shortages.

Read more about how the United Nations plans to combat desertification using the weblink in the Resources tab.

Resources

 Weblink United Nations Convention to Combat Desertification

1.6.9 How can people prevent or mitigate desertification?

DESERTIFICATION IN AFRICA

Located on the southern edge of the Sahara Desert in Africa, the Sahel is a semi-arid grassland/savanna biome that stretches from the Atlantic Ocean on the west coast to the Red Sea on the east coast. The Sahel includes parts of Sudan, Chad, Niger, Mali and southern Mauritania, and forms a buffer zone between the hot Sahara Desert and the humid savanna grasslands further south.

The Sahel has long been used for grazing and farming. Herders grazed their cattle, sheep and camels on the natural pasture grasses, thorny shrubs and baobab trees, and crops such as millet and groundnuts were grown when rainfall allowed. However, over-grazing and over-farming of this fragile region during the 20th century took its toll. With the depletion of vegetation, followed by several years of drought, the region was quickly consumed by the advancing sands of the Sahara. Relentless wind erosion and occasional water erosion led to excessive losses of topsoil, making the region more like a

FIGURE 1.46 Herders still graze their livestock in the Sahel, but available vegetation is decreasing.

desolate wasteland and largely uninhabitable. During the 1970s, an estimated 100 000 people starved, crops were wiped out and most cattle herds perished. Chad, one of the most central countries in Africa, lies astride the Sahel, and the length of its wet season can vary by as much as a third from year to year. Consequently, dry periods have a significant impact on the pastureland and on food security. The Sahel has entered into a negative feedback loop of increased desertification and increased climate change. The Sahel region is predicted to be an area most at risk of climate change in the future (see figure 1.47). Unless efforts are made to reduce the spread of high albedo land cover and increase ability for carbon sequestration, desertification will continue. Additionally, global changes to precipitation patterns and temperature patterns could occur, placing an ever increasing number of African countries at greater risk of **food insecurity**.

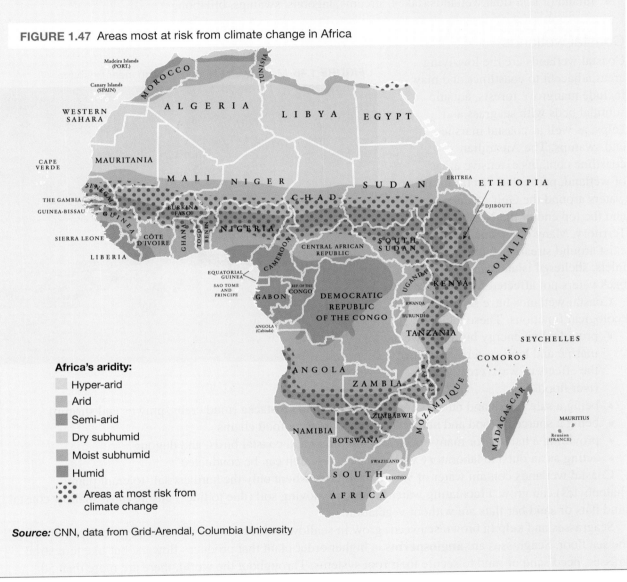

FIGURE 1.47 Areas most at risk from climate change in Africa

Africa's aridity:
- Hyper-arid
- Arid
- Semi-arid
- Dry subhumid
- Moist subhumid
- Humid
- Areas at most risk from climate change

Source: CNN, data from Grid-Arendal, Columbia University

1.6.10 Coastal biogeographic areas — mangroves and wetlands

The term 'wetlands' is used to describe low-lying areas containing salt marshes, swamps, bogs, peatlands and mangroves. A wetland is a discrete ecosystem that is regularly inundated by water, and so has its own unique plants and wildlife, particularly birds. Even though wetlands may be found in most countries, there are generally local variations depending upon climate, topography, soil-type, water quality and type, and even human activities.

Wetlands are usually classified into two groups.
- Coastal tidal wetlands (mangroves, salt marshes)
- Inland or non-tidal wetlands (lakes, streams, lagoons, swamps, billabongs)

Coastal wetlands

Coastal wetlands are the low-lying areas adjacent to coastlines and may include mangrove forests, aquatic subtidal beds with seagrass and kelps, as well as coastal marshes and swamps. The Australian coastline contains extensive areas of wetland, particularly in far northern waters around the Gulf of Carpentaria and the top end of the Northern Territory. Many other smaller pockets exist around stream estuaries, bays, inlets, sheltered islands and secluded backwaters not affected by **littoral drift**.

FIGURE 1.48 Mangrove at Nudgee Beach, near Brisbane

Source: Bill Dodd

Coastal wetlands have several key ecological functions. These include:
- providing an energy buffer between marine and land systems to reduce the effects of wave erosion and river flooding
- being a safe haven and nursery for juvenile fish and crustacea (mud crabs, prawns and shrimp)
- being a source of food and nutrients for many marine food chains
- providing a habitat for many species of migrating and coastal birds, and dugong
- acting as an outside laboratory where marine research can be conducted.

Coastal wetlands contain water of varying salinities where only the hardiest salt-tolerant plants (halophytes) can grow. Fluctuating water levels and moving soil (due to tides) ensure most shallow coastal mud flats or sand bar flats are without vegetation.

Seagrasses and kelp (a brown seaweed) grow in shallow coastal waters where sunlight can penetrate to the sea floor. Seagrasses are **angiosperms**, a higher order plant that produces flowers, but like terrestrial grasses, need sand or mud to secure their root systems. Throughout the world, there are more than 50 species of seagrass, 25 of which are found in Australia. Although they provide food and protection for small marine creatures, they are regarded as a nuisance to some boaters because they wrap around boat propellers.

The dominant coastal wetland plant is the mangrove, a tree with aerial root systems capable of growing in saline water and compressed sand that is low in oxygen (anaerobic). When undisturbed, mangroves become so dense they form forests. The thick network of intertwined roots is covered by saltwater at high tide and exposed at low tide. Mangrove root systems act like baffles, slowing water movement between tides and trapping silt. Mixing with detritus from the sea floor and algae, this sediment provides nutrients not only for the mangrove plants, but also the small crustacea and fish that seek protection from larger predators. Mangrove forests are among the most carbon rich forests, storing carbon dioxide and other greenhouses gases in their flooded soils for millennia. Despite the significant ecosystem service that they provide, mangroves are being destroyed at an alarming rate. Anthropogenic activities such as coastal development and **aquaculture** reduce the spread of mangroves, and scientists estimate that between 1980 and 2000, 35 per cent of global mangrove forest was lost. Consequently, due to changes in climate, there is an increase in extreme weather events, such as coastal low pressure systems, which are causing significant coastal erosion because the mangrove forests are no longer there to buffer the wave energy. As such, a negative feedback loop has been created.

Inland wetlands

Inland or non-tidal wetlands are shallow, freshwater areas not affected by tidal action and include lakes, streams, lagoons, marshes, swamps, billabongs and bogs. They are constantly inundated because the water table is either at or close to the surface. Inland wetlands are generally high in nutrients because little is removed from them. As a result, they support a variety of aquatic plants such as reeds, grasses, rushes, sedges, water lilies and wildflowers.

There are some basic differences between the wetland types. Marshes tend to support smaller plants such as sedges, reeds and grasses, while swamps are more nutrient-enriched and can support trees. A bog is a wetland where dead plant matter and moss accumulate to form peat, a high-carbon compound once dug up and used as a fuel.

Inland wetlands also perform several key ecological functions, including:
- providing important fish and bird habitat
- helping to mitigate and control erosion
- assisting with flood control and storm runoff
- use as recreational sites for people to enjoy
- use as nature reserves where biological studies are performed
- acting as carbon sinks.

Much like mangrove forests, wetlands are ecosystems that are at risk to anthropogenic activities, which threaten their resilience to changes in climate despite their ability to sequester carbon. Wetland vegetation absorbs carbon dioxide through photosynthesis but, because they are water logged, any decaying organic matter releases methane. As such, increased rates of methane in the atmosphere will increase temperatures, thus increasing the rate of decaying matter; a feedback loop is created. Additionally, changes to global precipitation and temperature patterns will either increase or decrease the spatial distribution of wetlands globally depending on their localised climatic changes.

Ramsar sites

In 1971, UNESCO drew up an inter-governmental convention to encourage countries to protect and manage their unique wetlands to protect wildlife and migrating birds. Because the first treaty was signed at Ramsar in Iran, it is now commonly called the Ramsar Convention. Globally there are 2331 Ramsar locations covering more than 2 million km^2. The UK has the most individual sites while Bolivia has the largest area of wetlands listed. Australia has 65 Ramsar sites covering an area of about 8.3 million ha (see figure 1.49).

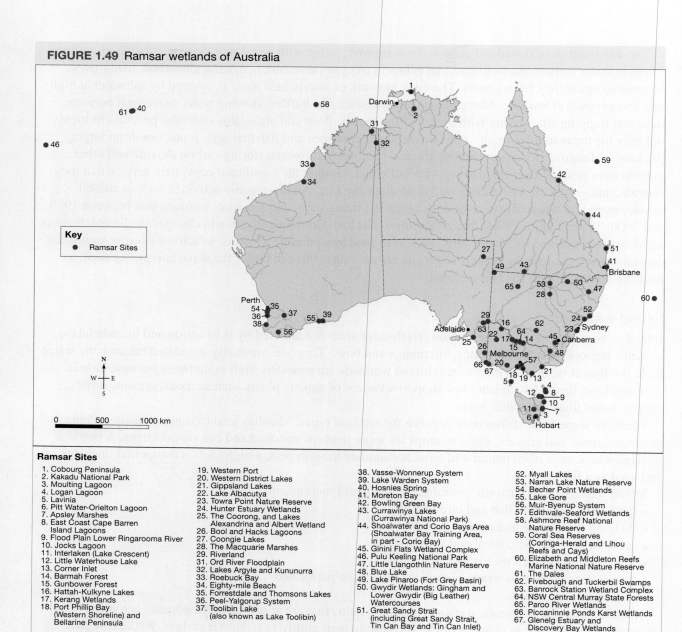

FIGURE 1.49 Ramsar wetlands of Australia

Ramsar Sites

1. Cobourg Peninsula
2. Kakadu National Park
3. Moulting Lagoon
4. Logan Lagoon
5. Lavinia
6. Pitt Water-Orielton Lagoon
7. Apsley Marshes
8. East Coast Cape Barren Island Lagoons
9. Flood Plain Lower Ringarooma River
10. Jocks Lagoon
11. Interlaken (Lake Crescent)
12. Little Waterhouse Lake
13. Corner Inlet
14. Barmah Forest
15. Gunbower Forest
16. Hattah-Kulkyne Lakes
17. Kerang Wetlands
18. Port Phillip Bay (Western Shoreline) and Bellarine Peninsula
19. Western Port
20. Western District Lakes
21. Gippsland Lakes
22. Lake Albacutya
23. Towra Point Nature Reserve
24. Hunter Estuary Wetlands
25. The Coorong, and Lakes Alexandrina and Albert Wetland
26. Bool and Hacks Lagoons
27. Coongie Lakes
28. The Macquarie Marshes
29. Riverland
30. Ord River Floodplain
31. Lakes Argyle and Kununurra
32. Roebuck Bay
33. Eighty-mile Beach
34. Forrestdale and Thomsons Lakes
35. Peel-Yalgorup System
36. Toolibin Lake (also known as Lake Toolibin)
37. Vasse-Wonnerup System
38. Lake Warden System
39. Hosnies Spring
40. Moreton Bay
41. Bowling Green Bay
42. Currawinya Lakes (Currawinya National Park)
43. Shoalwater and Corio Bays Area (Shoalwater Bay Training Area, in part - Corio Bay)
44. Ginini Flats Wetland Complex
45. Pulu Keeling National Park
46. Little Llangothlin Nature Reserve
47. Blue Lake
48. Lake Pinaroo (Fort Grey Basin)
49. Gwydir Wetlands: Gingham and Lower Gwydir (Big Leather) Watercourses
50. Great Sandy Strait (including Great Sandy Strait, Tin Can Bay and Tin Can Inlet)
51. Myall Lakes
52. Narran Lake Nature Reserve
53. Becher Point Wetlands
54. Lake Gore
55. Muir-Byenup System
56. Edithvale-Seaford Wetlands
57. Ashmore Reef National Nature Reserve
58. Coral Sea Reserves (Coringa-Herald and Lihou Reefs and Cays)
59. Elizabeth and Middleton Reefs Marine National Nature Reserve
60. The Dales
61. Fivebough and Tuckerbil Swamps
62. Banrock Station Wetland Complex
63. NSW Central Murray State Forests
64. Paroo River Wetlands
65. Piccaninnie Ponds Karst Wetlands
66. Glenelg Estuary and Discovery Bay Wetlands

Source: © Commonwealth of Australia 2019

COASTAL LAND RECLAMATION IN CHINA

During the 20th century, China moved rapidly from being a developing farming nation to one of the most powerful urban and industrial countries in the world. We have already seen that much of China is desert, so most change has occurred along the coastline. To meet this demand for port and shipping development, almost 800 000 km² of China's mangrove and tidal flat coastline has been reclaimed and converted into container ports, harbours and slabs of concrete for future expansion. Land reclamation is the process of constructing walls to keep the sea out and then filling the area with soil brought in from elsewhere.

FIGURE 1.50 Coastal land reclamation in China

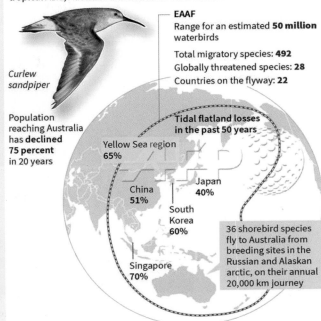

Source: © EAAFP Secretariat

As part of the **Five-Year Plan** for economic development, governments and business people believed unattractive mudflats could be filled in with soil and used for ports and terminals. Because of China's strict laws to protect arable soils and farming areas, coastal areas became a virtual development free-for-all. This would not only reduce some of the land shortage problem but provide industrial hubs to provide economic growth and employment.

Since 2003, satellite images have shown that about 14 000 km of coastline between southern China and South Korea have been modified and about 1.4 million ha of tidal mudflats reclaimed for industrial development. Reports also show that more than 70 per cent of the Yellow Sea mudflats and areas either side of the Yangtze River delta were destroyed. Intensive reclamation also occurred along the northern coastline to assist both agricultural and aquaculture interests.

FIGURE 1.51 Development on the coast of China

The unfortunate cost of this coastal transformation has been an alarming loss of coastal wetlands and mudflats, essential for the survival of marine creatures and many migratory bird species. These wetlands have also provided a protective buffer against storm surges, typhoons and tsunamis.

Refer to the weblinks in the Resources tab for more information.

Resources

Weblinks China's relentless campaign

Flying for your life

Great news for shorebirds

CHAPTER 1 Land cover transformations and climate change

Activity 1.6g: Using satellite images to measure land cover change
Explain and analyse satellite data

Examine figure 1.52, which shows land reclamation due to coastal development between 1995 and 2010, and then answer the following questions.

FIGURE 1.52 Coastal development in Shanghai, (a) 1995, (b) 2001, (c) 2005, (d) 2010

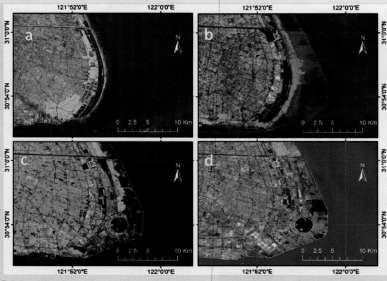

Source: Tian, Bo et al. Drivers, trends, and potential impacts of long-term coastal reclamation in China from 1985 to 2010, Figure 5 - The trends of coastal reclamation in China in the past three decades, *Estuarine, Coastal and Shelf Science*, Volume 170, 2016.

1. In a paragraph, describe what the coastline probably looked like in 1995 and how it might have been used.
2. Explain how the mangrove wetlands might have been removed and shallow waters filled in by 2001.
3. Describe what infrastructure and industrial projects would have replaced the original wetlands.
4. Explain how the reclaimed area might now be used for urban activities. Referring to figure 1.53, what do you think would have happened to coastal birds and aquatic wildlife?

FIGURE 1.53 The East Asian–Australasian flyway

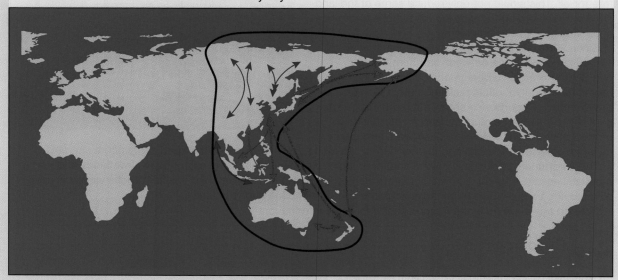

Source: © EAAFP Secretariat

1.7 Anthropogenic activity and global warming

1.7.1 Causes of global warming

Since the Industrial Revolution, there has been a significant change in the composition of the Earth's atmosphere. Centuries of clearing and burning forests, combined with an increased consumption of fossil fuels have given the atmosphere increased amounts of methane and nitrous oxide, as well as its highest known levels of CO_2 in the past 450 000 years. These are referred to as greenhouse gases because they trap infra-red radiation in the atmosphere and warm the Earth (like a greenhouse), and have been linked directly to global warming and probable climate change.

FIGURE 1.54 Methane bubbles up from the thawed permafrost at the bottom of the thermokarst lake through the ice at its surface.

Source: Katey Walter Anthony/ University of Alaska Fairbanks

The most common greenhouse gases are:
- carbon dioxide (CO_2) — although a naturally occurring gas, levels of CO_2 have increased due to burning fossil fuels like coal, oil and natural gas. We burn fossil fuels to generate electricity and use petrol to drive the transport industry and cars. In Australia, more than 70 per cent of electricity is derived from coal and 13 per cent from burning gas. The remainder comes from clean renewable energies such as hydro, solar and wind, which do not emit carbon. Half of the world's released carbon is absorbed by the ocean and biosphere, and the remainder accumulates in the troposphere at a rate of about 0.5 per cent per year.
- methane (CH_4) — a hydrocarbon found in natural gas, and produced by the burning of wood and fossil fuels, ruminant animals (cattle and sheep), rice paddies and landfills. Ice samples from Antarctica show methane levels are increasing by about 0.9 per cent every year. A major concern is the likelihood that huge volumes of methane currently trapped in rotting vegetation under the northern permafrost is escaping into the atmosphere as the frozen ground begins to thaw.
- chlorofluorocarbons (CFC-11, CFC-12) — gases used as refrigerator coolants, air conditioners, plastic foams and in solvents. Many used as aerosol propellants have been phased out since the **Montreal Protocol**.
- nitrous oxide (N_2O) — known as laughing gas, it is formed from the burning of fossil fuels, power plants, biomass and some fertilisers. When N_2O escapes into the atmosphere it has a life span of 150 years and is responsible for about three per cent of global warming.
- water vapour — also a greenhouse gas, but only has a tiny effect on the transfer of heat to and from the atmosphere. It has become an issue as increased CO_2 enables the atmosphere to hold more moisture. This means storms can be more violent and produce more rain.

Samples taken from Antarctic ice, frozen tundra lake mud sediments, annual rings of trees in old North American forests, and coral reefs reveal that the concentration of atmospheric CO_2 is now 40 per cent higher than it was two hundred years ago. According to the CSIRO, in pre-industrial times, CO_2 levels were approximately 200 ppm, increasing to between 280 and 300 ppm between some ice ages. Cape Grim in Tasmania has some of the cleanest air in the world but CO_2 readings taken there are around 395 ppm, while readings in Hawaii are constantly around 410 ppm.

In summary, the main causes of greenhouse gases are:
- burning fossil fuels for electricity and transport
- deforestation
- farming and agriculture.

Activity 1.7a: Carbon emitters
Interpreting information on a compound graph
1. Identify the countries shown to be the largest carbon emitters in figure 1.55 and find out why they put out so much carbon.
2. List the most polluting industries in each country and find out why they generate so much pollution.

FIGURE 1.55 The world's top greenhouse gas emitters

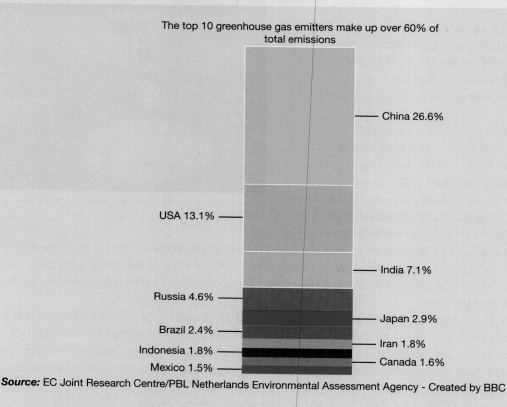

Source: EC Joint Research Centre/PBL Netherlands Environmental Assessment Agency - Created by BBC

1.7.2 Effects of global warming

Evidence shows that global warming is affecting environmental elements such as oceans and climate. The most significant effects are listed below.
- Thermal expansion of ocean water will raise sea levels about 20 cm by 2030. Low-lying coastlines like those in Bangladesh will become vulnerable to cyclones and storm surges, increasing the numbers of refugees. Islands like Kiribati, Vanuatu and the Maldives risk being inundated or having land degraded as salt water moves further inland. Mangroves may be drowned or move inland to areas now occupied by salt marshes.

- As oceans become warmer, the amount of water vapour in the air will increase, thus amplifying global warming. Melting of pack ice and glacial ice around the edge of Antarctica will also increase atmospheric moisture, but this may lead to higher than usual snowfalls in cold areas.
- Glaciers in temperate regions will melt more rapidly, reducing the extent of snow cover and causing snowfields to retreat. This will affect animals and plants in those areas.
- Changes in air and ocean temperature ultimately effect circulation patterns of air and water, inevitably leading to changes in weather and climate. If greenhouse gases continue to be produced at current rates, average global temperatures will increase by between 0.3 and 1.4 °C by 2030 and a further 0.6 to 3.8 °C by 2070. How maximum and minimum temperatures will be affected is unknown. A warmer atmosphere is also capable of holding more water vapour, thus increasing the chances of more storms and cyclones in coastal areas. Some inland areas are expected to get hotter and drier, while weather extremes such as El Niño will become more common in various parts of the world.

FIGURE 1.56 Global warming predictions

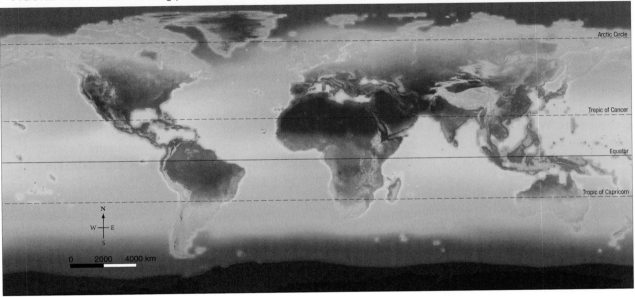

Source: NASA

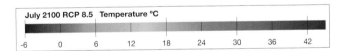

In Australia, global warming may cause:
- average temperatures to be 0.5 to 2.0 °C warmer within 200 km of the coast, and areas further inland could expect average temperatures to be 0.5 to 2.5 °C warmer
- evaporation to increase by between two and four per cent for each degree of temperature increase
- rainfall to increase by 10 to 20 per cent during normal wet periods but decrease by a further 10 per cent during drier months around coastal Australia. People could expect to receive more extreme wetter and drier seasons than normal.
- storms to be more intense and violent, often producing heavier falls of rain in shorter periods due to the increased water vapour. A good example of this was the Townsville deluge in January 2019, when more than 2000 mm of rain fell in just 12 days. Previously, Townsville's mean annual rainfall was about 1110 mm per year.
- wildlife numbers would be at risk, particularly species unable to adjust to warmer temperatures or a loss of food supply. Recent extreme events in Australia such as drought, fires and flooding have contributed to the reduction of many small animals and birds, and the prospect of warmer temperatures

is placing some site-specific creatures at risk. Examples include the mountain pygmy possum from the Kosciuszko region, the green ringtail possum from North Queensland, and coral communities in the Great Barrier Reef.

Global warming also affects food production and farmlands, forests (through stress and bushfires), wildlife, and the health of inland river systems (low flow rates and toxic algal blooms). Hot periods have become hotter and dry periods have become drier. Drought will affect grain and fruit crops, reducing yield quality or making them much more expensive. Cattle and sheep raising will be limited due to high temperatures and lack of water. Longer dry spells and more intense periods of rainfall may increase soil erosion, whereas higher levels of CO_2 would increase plant growth.

1.7.3 Climate change and its indicators

Climate change is a substantial variation from regular or expected climate patterns on a regional or global scale. The term is often attributed to increased amounts of water vapour, CO_2 and methane in the atmosphere due to changes caused by human factors as opposed to those resulting from the Earth's natural processes.

Although some people continue to debate whether climate change is even real, there is adequate evidence to show many features of the world's physical systems are warming. The changes in the atmosphere and oceans are the most noticeable, but changes in the land and soils are also noticeable to some extent as shown in figure 1.57. The Earth has passed through cooling and heating phases in the past, initiated by events such as changes in orbital position, ice ages or smaller random events such as the eruption of Krakatoa or meteorite collisions. However, the atmospheric and ocean trends of the past fifty years are conspicuous. While each one of these trends is not specifically 'climate change', collectively they are pointing to adjustments that affect climate in ways we have not seen before. According to the Australian Government's Department of the Environment and Energy, the most significant changes attributed to climate change are:

- record high surface air temperatures
- increased average number of hot days per year
- decreased average number of cold days per year
- increasing intensity and frequency of extreme events (e.g. fires, floods)
- changing rainfall patterns
- increasing sea surface temperatures
- rising sea levels and possible salt inundation of coastal areas
- increasing ocean heat content
- increasing ocean acidification
- changing Southern Ocean currents
- melting ice caps and glaciers
- decreasing Arctic sea ice.

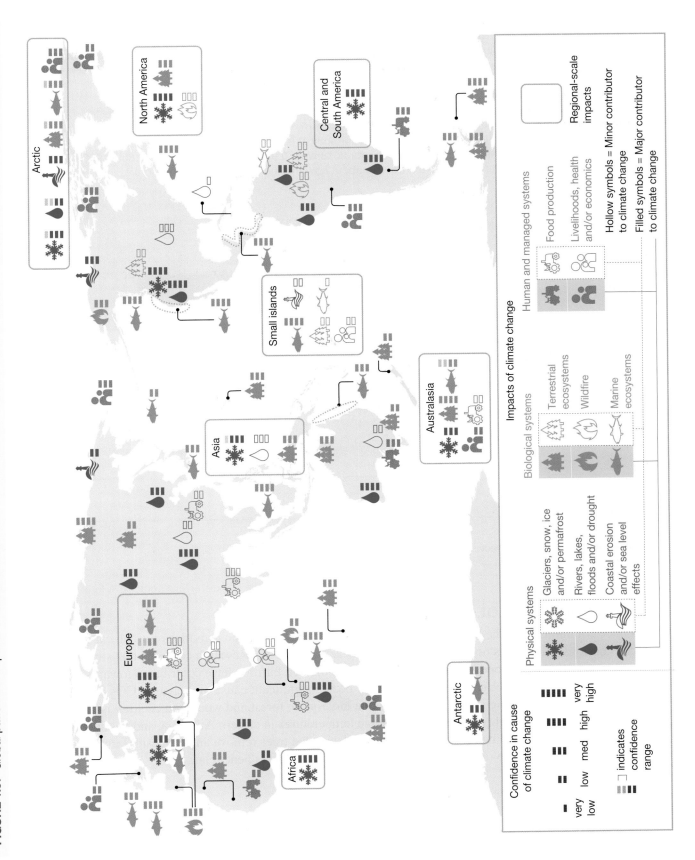

FIGURE 1.57 Global patterns of impacts in recent decades attributed to climate change

Activity 1.7b: Trends in global warming and climate change
Analyse and evaluate data

Choose six of the changes listed by the Australian Government's Department of the Environment and Energy and research data to support these claims. The weblink **Climate Reality Project** in the Resources tab can help with further research.

1. Analyse your data carefully and attempt to arrive at some possible effects.
2. Based on the data available to you, project possible future trends if people made little or no change to present behaviours.
3. Data shows that the world is getting warmer, and it currently at least one degree Celsius warmer than pre-industrial times. Global average temperatures for 2018 were 0.98 °C above the levels of 1850–1900. Using this information and figure 1.58, extrapolate about how this increase will impact on future communities in Australia.

FIGURE 1.58 Average projected warming by 2100

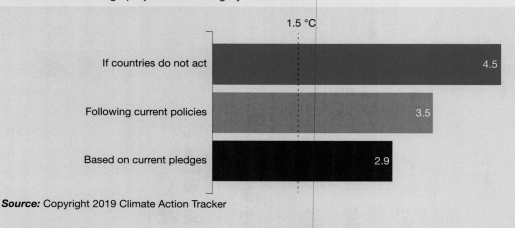

Source: Copyright 2019 Climate Action Tracker

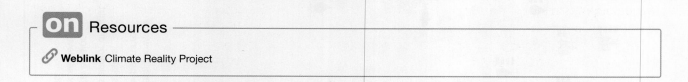

Resources

🔗 **Weblink** Climate Reality Project

1.7.4 The impacts of land cover change on climate

It was once common to put the blame for global warming on anthropogenic activities of urban car and factory emissions, but studies now show regional areas also contribute. Research confirms that since Australia was settled by Europeans, land clearing for grazing and farming have contributed towards Australia's dry spells and droughts, and has probably been a significant cause of increased CO_2 levels in the atmosphere and climate change. Data now shows that large-scale forest and woodland clearing is linked to intensity and extent of droughts, increases in average temperatures in eastern Australia, and a reduction in rain periods in most areas. When there is scientific evidence that trees remove and sequester large quantities of CO_2, the question must be raised as to why further forest clearing is beneficial.

FIGURE 1.59 Climate change amplification

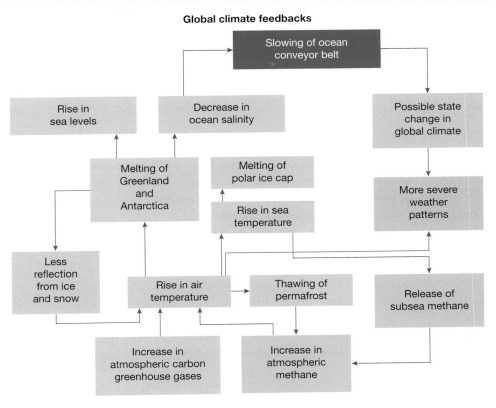

Because trees transpire moisture into the atmosphere as well as absorb CO_2, it is reasonable to assume a reduction of vegetation will have implications for the composition of both soil and air and eventually climatic conditions. Scientists believe that amounts of water vapour released into the atmosphere due to evapotranspiration are decreasing so the 'cooling effect' is lost. At the same time, heat radiated back into the atmosphere from the land is increasing because the dried-out soil surface is now hotter. These effects are magnified even further when periodic events such as El Niño occur. The most affected areas are in eastern Australia, between central Queensland and Victoria. Most alarming of all, is that once these events have occurred, they are almost irreversible.

For more information refer to the weblink **Land clearing and climate change in Queensland** in the Resources tab.

1.7.5 What can be done to mitigate climate change?

If weather can be forecast accurately a few days beforehand, is it possible to forecast climate change years in advance? Unfortunately, forecasting weather and climate changes are totally different challenges. Because weather is about short-term (daily) changes to the atmosphere, its immediate alterations are calculated on impending variables that can be measured, observed and displayed on a **synoptic chart**. It is like being able to predict traffic congestion at peak hour or following a road incident.

Predicting climate change is like estimating the flow of the whole road network over a decade and would include many unknown variables, such as improvements to public transport, future road budgets, introduction of driverless cars, and fuel and ticketing costs. Forecasting climate change requires consideration of many long-term factors such as rainfall reliability, drought, atmospheric and ocean warming, planetary wind circulation patterns, levels of CO_2 and methane emissions, methods of electricity generation and energy consumption, and many more. However, once human behaviours become established as a consistent pattern, it is possible to extrapolate data and predict conceivable scenarios and effects of climate change.

What are the future implications for people and environment?

With prospect of further climate change likely, it is expected that storms, cyclones, droughts and bushfires will become more severe across the Australian landscape. As more forests (carbon sinks) are lost to land clearing and fire, the atmosphere will inevitably increase its levels of CO_2.

At a global level, countries must do whatever is possible to reduce global warming. The most obvious strategies are to reduce further deforestation and rehabilitate damaged areas by planting trees as well as re-think energy production dependent on carbon fuels. Because these strategies have huge capital costs, governments need to take the lead and invest money into such changes.

At a community level, we need to:
- investigate ways of removing carbon dioxide from factories and cars, and either storing it underground or in the oceans
- encourage landowners to plant more trees or develop agroforestry on unproductive land
- provide incentives such as taxation relief to industries making genuine progress at lowering carbon emissions
- reduce land clearing and planting more trees — research has shown that planting saltbush in semi-arid regions will absorb up to 20 tonnes of carbon per hectare after only three years.

By relying on fossil fuels as our major energy source, Australia is missing out on opportunities for developing alternative energy solutions. We are also at risk of being left behind when industrial and business reform occurs in order to meet the new energy technologies. Australia's energy and greenhouse policies need to be separated from political decision making, a process which sometimes only looks as far ahead as the next election.

At a personal level, it is important that individuals see issues like global warming and climate change from a global point of view and appreciate that we are all part of the Earth's physical systems. Strategies for individuals to help save the planet from climate change include:
- making genuine efforts to save energy around the home or school and reducing demands on power stations — turning off or using fewer lights, air conditioners and appliances
- purchasing appliances with a high energy rating
- using more renewable forms of energy such as wind, solar and water power to lower coal and oil consumption
- walking, cycling, using public transport or participating in car pools rather than using the family car for all travel
- recycling and reducing demand for landfill.

While large amounts of money are needed to repair many of our old environmental problems, there is also a need to improve future planning and prevent mistakes happening. Successful proactive planning for the future can only occur by understanding today's environmental issues, appreciating why they exist, and confronting them with intelligent solutions. Tomorrow's decision-makers and practitioners will need to make wise choices and have the resolve to succeed.

For more information and resources about climate change refer to the weblinks in the Resources tab.

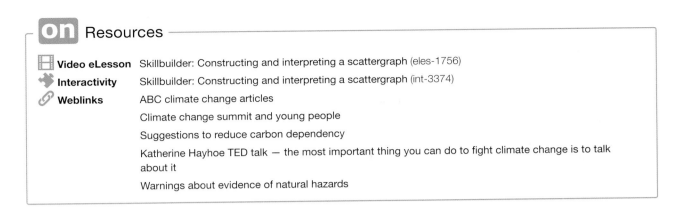

Treaties, protocols and global conferences

Greenhouse gas emissions had been on the radar for forty years when CO_2 levels began to increase. In 1987 several industrial countries, including Australia, signed the Montreal Protocol, an agreement to begin phasing out greenhouse gases. At first, it was assumed that only the **developed world** produced greenhouse gases, but as more developing countries turned to fossil fuels, a broader agreement was needed. At the Earth Summit in Rio de Janeiro in 1992, 154 countries pledged to reduce greenhouse gases to levels lower than in 1990 by the year 2000. Only Germany and Great Britain achieved this target.

Since 1995, the **United Nations Climate Change Conferences (UNFCCC)** have been held annually to promote a common approach to cleaner air and global warming. In 1997, 84 countries agreed to sign the Kyoto Protocol in Japan, and commit to cut emissions of six greenhouse gases so they would be 5.2 per cent lower than 1990 levels by the end of 2012.

Despite the evidence and pledges, many countries found it difficult to meet their carbon reduction targets. Financial crises, domestic politics and fears of energy shortages became barriers to good intention and saving the planet. Individuals including David Attenborough and Al Gore (who won a Nobel Prize for his documentary about the environment and climate change *An Inconvenient Truth*), as well as organisations like the **Intergovernmental Panel on Climate Change (IPCC)** have struggled to deliver their messages about sea level rising and climate change, despite the scientific evidence. International forums and conferences have resisted or been unable to reach agreement on carbon reduction. Some countries, including Australia and Norway, have been even allowed to increase emissions, Australia by eight per cent and Norway by one per cent. Australia used Article 3.7 of the protocol to gain carbon credits for programs that might reduce tree loss. Australia also gained favourable target concessions due to our dependence on fossil fuel exports. In retrospect, it is doubtful if these modifications were genuine or would even work, although some improvements have been made in agricultural practices like energy conservation, agroforestry, reduced biomass burning and improved manure management. Development of renewable energy programs using photovoltaic cells, wind turbines and hydroelectric plants, particularly in remote areas, has also continued.

Several global climate conferences have closed without clear progress because some world leaders have not been prepared to make genuine concessions and have been more concerned about their own domestic political situations. This has made it difficult to seek common ground and negotiate practical solutions. The 2009 UNFCCC summit in Copenhagen closed with vague promises to assist developing countries but few nations could settle on future targets. More recently, delegates at the 2015 conference in Paris agreed to keep the global temperature rise this century 'well below' two degrees Celsius and to make serious efforts to aim at a 1.5 °C increase. Since then, countries have pulled out of these agreements. More information about the US withdrawal from the Paris Agreement can be found in the Resources tab.

The most recent Climate Change Conference was COP 24, held at Katowice, Poland in 2018. Once again, there were huge divisions between what countries were prepared to do. The UN warned world leaders that pledges made at Paris in 2015 were at risk of failing because countries were not adhering to some of the difficult goals. IPCC scientists argued that some emission targets did not go far enough and that if they were not kept below 1.5 °C, the world was in for a very grim future. There was also concern about different rules for both developing and developed countries, and who was going to monitor adherence to targets, particularly for the biggest carbon emitters: the US and China.

Fifteen-year-old Swedish climate activist Greta Thunberg spoke in the closing stages of the 2018 conference. She wanted to know why, when climate change became news 30 years ago, world leaders did nothing to fix it. Thunberg also made a plea for delegates to take the situation more seriously, and blamed the present situation on a lack of commitment and care from politicians by saying, 'The year 2078, I will celebrate my 75th birthday. If I have children, maybe they will spend that day with me. Maybe they will ask me about you. Maybe they will ask why you didn't do anything while there still was time to act.'

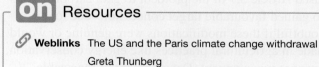

FIGURE 1.60 Greta Thunberg speaking at a school strike for climate change in Sweden.

You can watch Thunberg's speech and read more about children's opinions on the actions taken to mitigate climate change using the links in the Resources tab.

on Resources

🔗 **Weblinks** The US and the Paris climate change withdrawal
Greta Thunberg
Children are telling us we have failed them

Activity 1.7c: Describing human responses to climate change
Explain and analyse different responses
1. Watch the video speech of Greta Thunberg at COP 24 in the Resource tab and list the main points of her address.
2. Explain why many young people are frustrated with the lack of action by political leaders, given Katowice, Poland was the 24th climate meeting.
3. How effective was Thunberg's message? Has there been any response?
4. Read the article in the following information box. It first appeared in a Queensland Senior Geography textbook more than 25 years ago. Why do you think very little has been done to correct this trend?
5. What parts of the article are consistent with what is happening today?
6. How does this text align with the message delivered by Thunberg?

An important area of scientific geography undergoing change is climatology. Since atmospheric scientists have established that the greenhouse effect and ozone depletion are continually changing the composition of the air, climatologists now have the difficult task of attempting to predict how these changes may affect weather and climate circumstances around the world. As increased quantities of carbon dioxide, methane, chlorine and water vapour cause variations to usual temperatures, rainfall and evaporation rates, long term climate changes seem inevitable. However, what will be the extent of these changes and how might people adjust to them?

Climate changes consistent with greenhouse forecasts have already started taking place in Australia, although it has taken extreme circumstances of drought (1982–3 and 1991–2) in eastern Australia and flood (1983 and 1992 in south-east Queensland) to make the community realise that things are different from normal. Nevertheless, mean temperatures all around the globe have been steadily increasing, with the 1980s including eight of the world's warmest years since records began in the 1850s.

Source: Bill Dodd, *Spectrum: Geographical Perspectives on People and their Environment*, p 208, Jacaranda, 1994, based on data provided by CSIRO.

1.7.6 Effects of climate change on land cover

Climate change creates challenges for the future **sustainability** of natural environments. The interconnections that occur in global climatic systems with the changing of climate significantly impact the physical systems that have created various land covers. In the following case studies, various land covers will be investigated to analyse the interconnections between climate change and the viability of achieving sustainability for these land covers.

Activity 1.7d: Urban growth and climate change

Explain and analyse data

Refer to figure 1.61, which shows large cities and their vulnerability to climate change, to answer the following questions.
1. Which continents have cities with the highest and lowest levels of risk to climate change?
2. Explain why there are differences.
3. Identify where Australia is positioned. Explain its relative position in terms of climate vulnerability.
4. Find out why Kinshasa and Lagos are mentioned specifically.
5. Why do you think European cities may be less vulnerable than most other places?

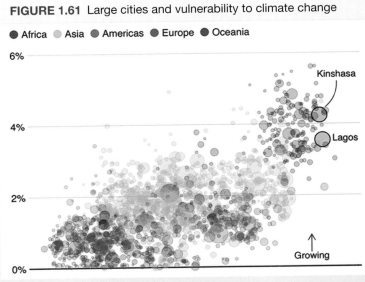

FIGURE 1.61 Large cities and vulnerability to climate change

Source: Verisk Maplecroft - Created by BBC

THE GREAT BARRIER REEF

The Great Barrier Reef, stretching from eight degrees south to 24 degrees south along the coastline of Queensland, is a unique ecosystem that is at risk to climate change. In some sections the reef is more than 200 km offshore, while near the mouth of the Daintree River it abuts local beaches and rainforest. Separating the reef from the mainland, is a narrow shallow channel of water that is largely protected from prevailing winds that cross the open Pacific Ocean. It is difficult for large ships to navigate close to the coast due to the thousands of small land islands and coral atolls scattered throughout the channel.

Very specific biophysical interactions need to occur to allow coral reefs to form. Optimal coral growth requires:
- clear, shallow water
- temperatures between 21 and 29 °C
- relatively stable saline water.

Any changes to these conditions alter the ability for the coral polyps and zooxanthellae, the organism that lives inside the coral polyps, to survive. Anthropogenic activities are altering these conditions for coral through direct activities such as agriculture and pollution, and indirect activities such as those that have led to increases in sea level, sea surface temperature and severe weather events.

Climate change has led to sea level rises, which increase the depth of coral from the sea surface, consequently altering the temperatures of the areas where coral live. The speed at which this is occurring is happening at a rate much faster than the corals' ability to build up the bommie, because such coral growth is stunted.

Increases in sea surface temperatures have meant that the ocean is becoming too hot for coral to grow, which results in large-scale bleaching events. Coral bleaching has been linked to increasing turbid water from farming run off.

With the coral reef system already weakened by bleaching or made vulnerable by changes to optimal coral growth, the increased rate of severe weather events have reduced the reefs resilience.

The biodiversity of this marine system is remarkable: it includes about 2900 individual reefs, approximately 1500 fish species, 400 coral species, 4000 mollusc species, 242 bird species, and a large range of crustaceans, sponges, anemones and worms. If the coral reef can no longer withstand the forces of climate change, the biodiversity of this ecosystem will diminish.

Today, management of the reef is based on the following principles.
- The ecosystem must be protected and passed on to future generations for the benefit of all people.
- Use of the reef must be reasonable and sustainable, while still allowing people to enjoy it.
- Public and community participation is essential.
- Monitoring and evaluation of programs must occur.

Under the terms of the GBRMP Zoning Plan (2003), a number of special purpose zones were established to help manage the reef. These were:
- Preservation Zone, where the public is not allowed access such as Jacqueline Reef in the Whitsunday Island Group.
- Marine National Park Zone, where fishing is not permitted and only low impact activities like swimming, snorkelling, sailing and boating allowed.
- Scientific Research Zone, where only marine research is permitted, eg Australian Institute of Marine Science near Townsville and Bunker Group near Gladstone and limited research at Green Island, off Cairns.
- Buffer Zone, which identifies protected areas in their natural condition. Limited boating and pelagic fishing for marlin, is allowed but line and spear fishing are prohibited.
- Conservation Park Zone, where public may access for general recreational use and limited fishing such as Magnetic Island (Townsville) and Hamilton Island (Proserpine). Regular ferry and shipping routes are permitted with limitations.
- Habitat Protection Zone, where sensitive marine habitats have been identified and trawling is prohibited.
- General Use Zone, where most 'reasonable' activities are allowed without permission, including shipping and trawling.

Source: © Great Barrier Reef Marine Park Authority

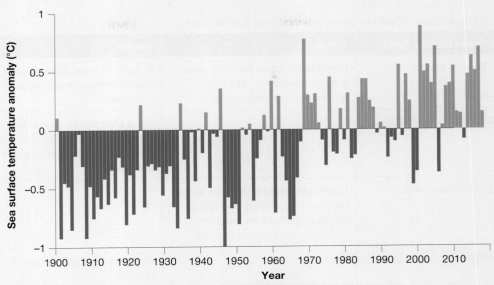

FIGURE 1.62 Sea surface temperature anomalies for the Great Barrier Reef, 1900–2018

Source: © Copyright Commonwealth of Australia 2019, Bureau of Meteorology

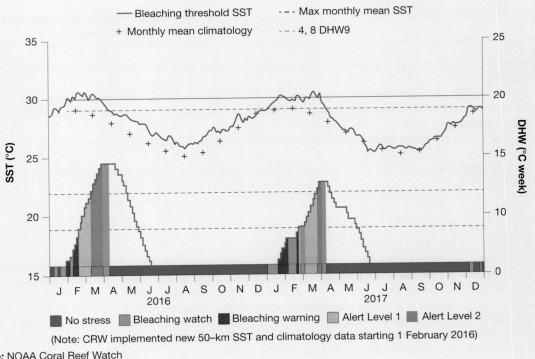

FIGURE 1.63 Link between sea surface temperature and coral bleaching events at Lizard Island, 2016–18

(Note: CRW implemented new 50–km SST and climatology data starting 1 February 2016)

Source: NOAA Coral Reef Watch

Activity 1.7e: Reviewing the GBRMPA management plan
Communicate and justify your response

TABLE 1.7 Causes and effects of threats to the Great Barrier Reef

Threat to Great Barrier Reef	Effect	Cause
Crown-of-Thorns starfish	Preys upon coral and kills large areas of reef when starfish numbers increase rapidly	Scientists think outbreaks of starfish coincide with nutrient increases
Sedimentation from mainland run-off		Land clearing for ports, farming and other construction projects near waterways that empty into the ocean
Acidification of ocean water	Lower pH affects ability of some creatures to survive and has a damaging effect on creatures with calcium shells and coral	
Coral bleaching		Ocean warming and climate change
Tropical cyclones and severe storms		Increased severity of cyclones due to climate change
General pollution and litter (plastics)	Kills marine creatures; ruins aesthetic effects of reef environment	
Oil spills and bilge dumping		Illegal navigation or shortcutting through shipping lanes

1. Study the information in table 1.7. What do you consider to be the greatest threat to the integrity and survival of the reef?
2. Write an extended paragraph explaining what the reef will possibly look like in another 20 years if no intervention is made to the cause of the threat. Support your response with either statistical or anecdotal evidence if possible.
3. Use figures 1.62 and 1.63 to explain the interconnection between changes in climate and optimal coral growth.
4. Extrapolate these interconnections to generalise about the impact that this would have on people, places and environments.

EFFECTS OF CLIMATE CHANGE ON ANTARCTICA

Antarctica is one of the most unique places on Earth. Its large continental land mass (about 1.5 times larger than Australia), and position over the South Pole makes it the coldest, driest and windiest continent in the world. It is colder than the northern Arctic because of its continental landmass and high elevation. Differences in latitude, altitude and local conditions also cause inland temperatures to decrease markedly from those of coastal areas. The coldest air temperature ever recorded was -89.6 °C in 1983 at Vostok, a Russian Antarctic base 1300 km from the coast and 3488 m above sea level. Temperatures in coastal Antarctica tend to range from summer maxima around 9.0 °C to winter minima of around -40 °C.

Antarctica's low annual precipitation qualifies it as a desert. On the interior plateau, annual snowfall is equivalent to 50 mm of rain, which is about half the amount received at Birdsville in the Simpson Desert. Light rainfall does occur near the coast but only in summer. Cyclonic gale-force winds are generated by the deep low pressure systems that form over the southern oceans. They also create a chill factor that sends temperatures lower than usual. With speeds that are often over 100 km/h for days at a time, and maximum gusts of up to

250 km/h, winds continually affect comfort and bring human activity to a standstill. The most severe gales and snowstorms tend to occur during winter months. Despite this harsh setting, evidence is emerging that climate change is now having a negative impact on the Antarctic environment and its ecosystems.

Scientists have recorded significant warming of both air temperatures and ocean temperatures, particularly in western Antarctica. Some scientists say that this region is warming faster than any other part of Earth. Recent findings include:

- warmer air temperatures over land/ice areas
- warmer seawater temperatures in the upper layers of the Antarctic Circumpolar current, a change greater than one degree Celsius since the 1950s
- irregular fluctuations of off-shore sea ice, which can double the cover of ice during winter and thus affect albedo
- increased melting of the Antarctic ice sheet, which covers 97 per cent of the continent at an average depth of approximately 2 km
- gradual retreating of some glaciers.

Because most discussion on climate change in the media is about where we live, remote places like Antarctica tend to be ignored by the public. As well, the sheer size of the land and ice mass means that effects of climate change are not occurring uniformly. For example, some coastal areas are experiencing increases in sea ice while in other places it is decreasing. This has implications for much of the marine wildlife. However, there is no escaping the fact that Antarctica is an enormous **heat sink**, which means it is absorbing excess heat. This has implications for the whole planet.

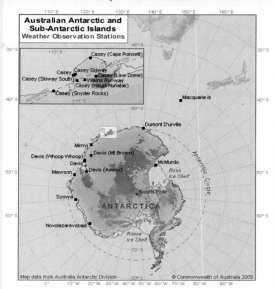

FIGURE 1.64 Map of Antarctica

Source: © Commonwealth of Australia, Bureau of Meteorology

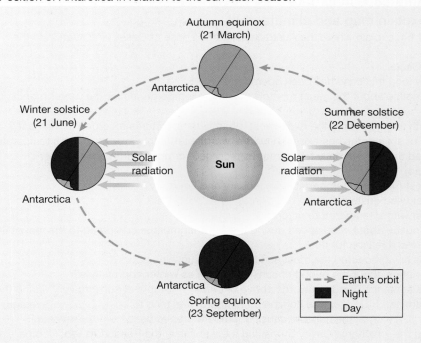

FIGURE 1.65 Position of Antarctica in relation to the sun each season

TABLE 1.8 Weather data for Antarctic base Casey 66.3 °S latitude, 110.5 °E longitude; 40 m asl, operational since 1989

Statistics	Jan	Feb	Mar	Apr	May	Jun	Jul	Aug	Sep	Oct	Nov	Dec	Annual
Mean maximum temperature (°C)	2.3	-0.1	-4.2	-7.6	-10.8	-10.7	-10.6	-10.8	-9.9	-8.0	-2.4	1.5	-5.9
Highest temperature (°C)	9.2	6.6	4.0	3.0	4.5	4.2	3.0	5.0	3.9	1.5	4.9	8.0	9.2
Year	1991	1991	2014	1990	2008	2013	2018	2007	1989	2014	1997	2005	1991
Lowest temperature (°C)	-10.0	-18.0	-25.1	-31.3	-34.4	-34.1	-34.2	-37.5	-33.8	-31.2	-23.4	-13.0	-37.5
Year	2012	2012	2014	1999	1994	1996	2016	2005	2017	1992	1995	1994	2005
Mean daily sunshine hrs	5.3	4.5	3.2	2.0	0.7	0.1	0.4	1.5	3.0	4.5	5.8	5.9	3.2
Max wind gust km/hr	163	187	241	223	215	217	241	221	241	213	184	181	
Year	2002	2014	1992	1991	1990	1993	1989	2003	2003	1992	1996	1995	2003

Source: © Commonwealth of Australia, Bureau of Meteorology

Activity 1.7f: Antarctica map data
Analyse and explain map and climate data

1. Using figure 1.64, explain where the Australian Antarctic base of Casey is in relation to the
   (a) coast line
   (b) Antarctic Circle.
2. Research how long this base has been operational.
3. What do you notice about the mean monthly maximum temperatures for the winter months of June, July and August?
4. In which month and year were the two lowest monthly temperatures recorded at the base?
5. Which period of six months has experienced the warmest monthly average (mean) temperatures?
6. Which year had the highest daily temperatures recorded for both March and October?
7. Which year had:
   (a) the warmest maximum temperature in January and February?
   (b) the highest wind gusts?
   (c) the coldest winter temperature recorded?
8. What do you notice about wind speed during the summer months compared to the rest of the year? Try to find out if there is a reason for this.
9. Which months have recorded the most and least daily sunshine hours?
10. What do you notice about the period of sunshine hours as winter approaches?
11. Complete this sentence using the words *above* or *below* — 'Hours of sunlight vary during the year. In summer, the sun is ... ... the horizon so temperatures tend to be ... ... average. During winter, the sun stays ... ... the horizon, causing temperatures to be ... ... average.
12. Have a look at the weather in Casey by visiting the weblink in the Resources tab. Compare some of this data to another Australia base such as Mawson, which has been operational since 1954. Make a list of similarities and differences.

Resources

Weblinks Casey
ASOC
Discovering Antartica

Activity 1.7g: Using remotely-sensed evidence
Explain, analyse and extrapolate future scenarios of land cover based on remotely sensed evidence

Read the article about Antarctic's loss of ice using the **Antarctica lost ice** weblink, found in the Resources tab.
1. What changes to the Antarctic ice cover have been exposed by satellite observations?
2. How has the loss of ice been interpreted in terms of future sea levels? Provide data.
3. This research has been translated into two alarming future narratives. What are they?
4. Why are they described as 'alarming'?
5. Extrapolate how these projected scenarios will impact on future marine ecosystems, climate and people.

FIGURE 1.66 Sea ice extent of Antarctica, January 2019

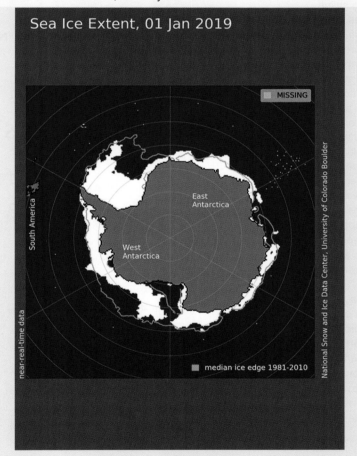

Source: Image courtesy of the National Snow and Ice Data Center, University of Colorado, Boulder.

Resources

Weblink Antarctica lost ice

THE TRAGEDY OF CLIMATE CHANGE IN BANGLADESH

Bangladesh, located at the head of the Bay of Bengal in southern Asia, is a small country extremely exposed to the effects of climate change. With a population of 165 million living in an area of 147 570 km^2 (about 60 per cent the size of Victoria), Bangladesh is one of the most densely populated countries in the world. Approximately 15 million people live in Dhaka, the capital city; the majority of the population lives along the low-lying coastline because of the fertile soils there.

Much of Bangladesh is positioned in a massive delta built by the Ganges, Brahmaputra and Meghna Rivers. Land is very flat and only a few metres above sea level. A large portion of the country literally floats on water, over which it has no control. With 57 streams running through the low fertile floodplains, Bangladesh depends on farming, making it one of the poorest countries in the world. The combination of high population density, poverty and potential for inundation (floods, typhoons, tidal surge) make Bangladesh one of the most vulnerable nations in the world in terms of global warming and sea level rising. A lack of emergency response resources also adds to the people's exposure.

Rice and maize farming occurs in **polders**, tracts of land enclosed by banks of earth built in the 1960s to keep out the sea and control irrigation. Polders have played a crucial role in preventing saline intrusion and tidal surges, but many are now in a state of disrepair, adding to Bangladesh's exposure.

Due to its location, Bangladesh had been forced to deal with extreme weather events, severe flooding and inundation from the sea. Even increasing meltwaters from glaciers thousands of kilometres away in the Himalayas, are a potential hazard for its farmers. Despite these challenges, Bangladesh has struggled through periods of extreme poverty and starvation, to a point where it now exports food, and has approximately 4 million women employed in garment manufacturing, although many of these have extremely poor working conditions.

Data shows that sea surface temperatures in the Bay of Bengal have increased more quickly than waters in many other parts of the world. This is mainly due to the warm climate and shallow depth of the bay. During cyclonic weather, storm surges can push walls of water more than 50 km into the delta river system.

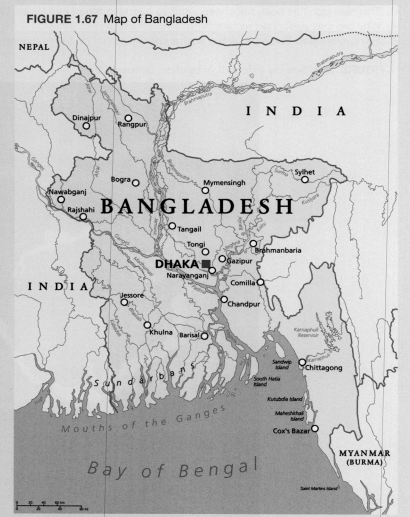

FIGURE 1.67 Map of Bangladesh

If sea levels in the bay rise as projected (up to 50 cm by 2050), the effects will be disastrous. Bangladesh will lose more than 10 per cent of its farmland through flooding, inundation and **salt intrusion**, and, should this occur, about 15 million people will be displaced. Some demographers predict that more than 200 000 coastal farmers will be jobless, rice and maize yields will fall, and the country will be forced to buy grains from overseas markets.

How can people adapt? In the short term, money needs to be spent improving the integrity of its polders, but long-term proposals might include:

- farmers in coastal regions changing from rice farming to aquaculture, especially shrimp farming, although this can be expensive
- investigation into planting crops more suited to dryland farming away from the coast.

FIGURE 1.68 Fields in Kantanagar near Dinajpur, Bangladesh

FIGURE 1.69 Rising sea levels in Bangladesh threaten houses.

Activity 1.7h: Discussion and decision-making

Evaluate and justify issues

Throughout this topic, much of the blame for increased levels of CO_2 in the air has been attributed to burning fossil fuels, land clearing and livestock farming. Consider the advantages and disadvantages of the strategies in table 1.9 as means of reducing levels of CO_2.

1. Examine each of these issues in terms of social, economic and environmental criteria, giving reasons for and against each one using a table like the following.
2. Decide which strategies may be most difficult to implement. Why?
3. Rank them in order based on possible acceptability, giving reasons for your decisions.

TABLE 1.9 Advantages and disadvantages of the strategies of reducing levels of CO_2

Strategy	Reasons for	Reasons against
1. Reduce land clearing in urban and rural areas.	Social: Economic: Environmental:	Social: Economic: Environmental:
2. Increase afforestation programs.	Social: Economic: Environmental:	Social: Economic: Environmental:
3. Reduce electricity generation from coal-fired power stations	Social: Economic: Environmental:	Social: Economic: Environmental:
4. Decrease livestock herds.	Social: Economic: Environmental:	Social: Economic: Environmental:

1.8 Review

1.8.1 Chapter summary

This topic has covered some key points about climate change and the transformation of global land cover.

What is land cover and its distribution?
- Land cover is the different materials that cover the earth. It refers to forest, grass, farmland, roads, building, exposed ground, lakes and water.

The Earth's physical systems
- There are four physical systems: the atmosphere, the lithosphere, the hydrosphere and the biosphere.
- These physical systems are connected by three energy cycles: the water cycle, the carbon cycle and the nitrogen cycle.

Global climate systems
- Our global climate systems are affected by wind patterns, heat transfer in the atmosphere and ocean currents.
- Global patterns of precipitation are closely aligned to wind patterns and heat transfer in the atmosphere.
- Heat energy is transferred between the equator and the poles by very large ocean currents, which are large masses of water that circulate water flow around the oceans.
- El Niño and La Niña are global events involving both the atmosphere and ocean waters, which affect weather and climate in Australia.
- The Earth's energy balance is the process of energy from the sun entering the Earth's system during daylight hours by radiation and warming the surface. The energy then leaves during the night hours by infrared radiation from the atmosphere.

- 'Albedo' is a term that means the proportion of light reflected by a surface. Light-coloured surfaces such as snow or ice reflect up to 95 per cent of solar energy so they have a high albedo. Dark areas like rainforest, ploughed soil or ocean water absorb most of the heat and reflect only small quantities away.
- Loss of landcover is affecting the ability of the surface to reflect or absorb heat, due to changes in the surface albedo.
- Greenhouse gases are a collection of naturally occurring gases in the troposphere that allow the sun's rays through and trap some of the heat. This prevents the Earth becoming too cold to be habitable. However, too much greenhouse gas raises the temperature of the Earth.
- The Indian Ocean Dipole is a measure of sea surface temperatures, in which alternating warm and cool ocean temperatures affect the rising and falling of atmospheric moisture. These give an indication of when dry or wet spells may occur over the western half of Australia
- The Arctic Oscillation and the polar vortex are two phenomena that play key roles in climatic patterns and the distribution of air.

Changes to land cover
- Biomes are very large regions of the Earth where area-specific plants and animals have adapted to the climatic conditions, soil and relief of that environment. For example, deserts and rainforests are biomes.
- Anthropogenic biomes are areas that have experienced sustained human interaction and has transformed land cover.
- Increased population growth and demand for food, water and other resources have increased the pressure on the land and fertile soils.

Anthropogenic activity and how it has transformed land cover
- Many human projects, in order to secure economic development, food security and employment, have resulted in long-term changes to the Earth's biophysical systems.
- Technology and increasing affluence have influenced the transformation of land cover.
- Forests cover about 31 per cent of the land surface, contain the highest levels of terrestrial biodiversity and up to 80 per cent of the total plant biomass. However, the amount of forests are dramatically declining.
- Deforestation refers to the intentional clearing of forests for other purposes such as farming, cattle grazing, logging, mining, and urban and industrial development, causing transformation of land cover.
- 'Rangelands' is a term used to describe grasslands, shrublands, woodlands, wetlands, and deserts that are grazed by domestic livestock or wild animals. They refer to the use of land for a particular human activity, whereas grasslands or savanna are terms that describe the climate and vegetation.
- The Earth contains large areas of hot, dry land known as desert. These are areas that receive less than 250 mm of rain in a year and where less than 50 per cent of the ground surface is covered by vegetation.
- Desertification is a form of land degradation where land that was once fertile and arable has been transformed into unproductive arid land, often due to human causes (overgrazing, vegetation removal) or natural causes (drought).
- Wetlands and mangrove forests are ecosystems that are at risk to anthropogenic activities, which threaten their resilience to changes in climate despite their ability to sequester carbon.

Anthropogenic activity and global warming
- Forests act as a carbon sink that absorb carbon dioxide and other greenhouse gases that would otherwise remain free in the atmosphere. When large areas of forest are destroyed, this vital role as a sink is lost.
- Greenhouse gases have been linked directly to global warming and probable climate change.
- There is adequate evidence to show many features of the world's physical systems are warming. These include trends in changes in rainfall, temperature, sea surface temperatures, ocean currents and melting ice caps.

- As well as urban car use and factory emissions, land clearing for grazing and farming have contributed towards Australia's dry spells and droughts and has probably been a significant cause of increased CO_2 levels in the atmosphere and climate change.
- Various global conferences and treaties have attempted to combat climate change. However, despite the evidence and pledges, many countries found it difficult to meet their carbon reduction targets.
- Actions can be taken at a community level to help combat climate change.

1.8.2 Key questions revisited

You should now be able to answer the following questions.
- What is land cover and its distribution?
- What processes connect the Earth's physical systems and affect land cover?
- What are the different types of land cover? (vegetation biomes, biogeographic areas, anthropogenic biomes)
- How does population growth, an increase in affluence and technology impact upon land cover?
- How do human activities like settlements, croplands, rangelands and forestry transform land cover surfaces?
- How do these transformations impact upon the Earth's systems?
- What are global climatic systems?
- What is climate change? How does it impact on land cover types?
- What are the implications of climate change on people and the environment and how might people best respond to them?

1.8.3 Practice Assessment 1

Go to the Resources tab for a practice assessment for this topic, along with a stimulus sheet and marking guide.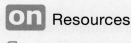

> **Resources**
>
> **Digital documents** Practice Assessment 1 Stimulus sheet (doc-31435)
> Practice Assessment 1 Examination: Combination response (doc-31433)
> Practice Assessment 1 Marking guide (doc-31434)

2 Responding to local land cover transformations

2.1 Overview

2.1.1 Introduction

The world around us is in a constant state of change: night changes into day, summer into autumn, forests become housing estates, roads carve their way across the land, seeds grow into giant trees and a dry stream bed can become a raging torrent after rain. All of these changes are felt most prominently at the local scale, in the places closest to us that we see, feel, hear and interact with.

Land cover changes are visible and prominent, and they are driven by either natural or human processes. Natural causes of land cover change include erosion, deposition and natural hazards while human causes include a local residential development, removal of native **vegetation** for farming, laying of roads, building an airport, creating a dump, growing plantation forests and mining for ore.

It is important to understand and manage land cover changes to ensure the balance of **development** and **sustainability**, and the retention of the features of local places that make them special. In this chapter you will learn strategies to manage land cover change and you will explore options for your fieldwork activity.

FIGURE 2.1 Natural land cover looking towards Hinchinbrook Island, Queensland

2.1.2 Key questions

- What is land cover change?
- How is land cover important to us environmentally, socially and culturally?
- What processes are involved in local land cover changes?
- What are the impacts of local land cover changes?
- How can we manage local land cover changes to maximise sustainability?

2.2 Processes resulting in land cover change

2.2.1 Anthropogenic processes

Anthropogenic, or human-induced, landform change has been occurring for as long as humans have existed. As the global population has increased and our technical capability as a species has progressed, our ability to impact on the environment has rapidly increased, particularly over the last century.

Humans also impact on land cover change by removing and moving sediment through erosion. When vegetation is cleared, usually for development, sediment is exposed and can be stripped away or removed. Water and wind are the two main drivers of erosion.

Two major anthropogenic processes are **urbanisation** and resource exploitation.

Urbanisation

Urbanisation is the movement of people into urban areas often from rural and remote areas. It is the growth in proportion of a population living in urban environments. This phenomenon is occurring across the world, with more people and a greater proportion of the global population living in cities. Currently, 54 per cent of people on Earth live in an urban place and this proportion has been steadily increasing over the last century. The United Nations (UN) predicts that, from 2015 to 2030, the global urban population will grow between 1.44 per cent and 1.84 per cent per year. Rates of urbanisation are highest in developing countries, with African and South American cities predicted to see the highest rates of urbanisation in the near future.

A high proportion of the Australian population has always lived in cities. European settlements that began as towns grew, attracting more and more people. Some people would move into rural areas as the country was opened up to farming but the trend over the past 100 years has clearly been one of increasing urbanisation. Figure 2.2 shows the components of population change in Queensland since 1981 and how this is projected to continue.

> Queensland's population change is comprised of three components: natural increase (births minus deaths), net interstate migration (interstate arrivals minus interstate departures), and net overseas migration (overseas arrivals minus overseas departures).

Resource exploitation

Our knowledge and understanding of how to use different **resources** is what has driven every technological advancement that has changed humanity, from the domestication of animals, to the development of metals, the invention of the printing press, to the current digital revolution. The use of resources by humans has changed the world fundamentally.

Natural resources are features of the natural environment from which we derive benefits, such as sunlight, **nutrients**, water, plant and animal species, and **minerals**. Almost everything seen in nature can be used benefit humans. All animal species, including humans, consume resources to survive, mainly in the form of other animals or plant species.

In a geographical context, natural resources are consumable and humans use resources to create materials that can sustain us or materials for our convenience, enjoyment and/or pleasure. Some broad examples of the use of resources include:

- **crops** and some domesticated animals for food and clothing
- minerals and **fossil fuels** to create plastics
- minerals and plants to create building materials like cement, wood and plastics that are used to build homes, cities, towns and businesses
- minerals and fossil fuels for vehicles and fuel for transportation
- minerals and fossil fuels to create many modern electronics
- minerals and fossil fuels as a source of power for homes and businesses
- renewable energy (solar, wind, geothermal, wave and hydro) to power homes and businesses.

FIGURE 2.2 Queensland government population projections to 2041

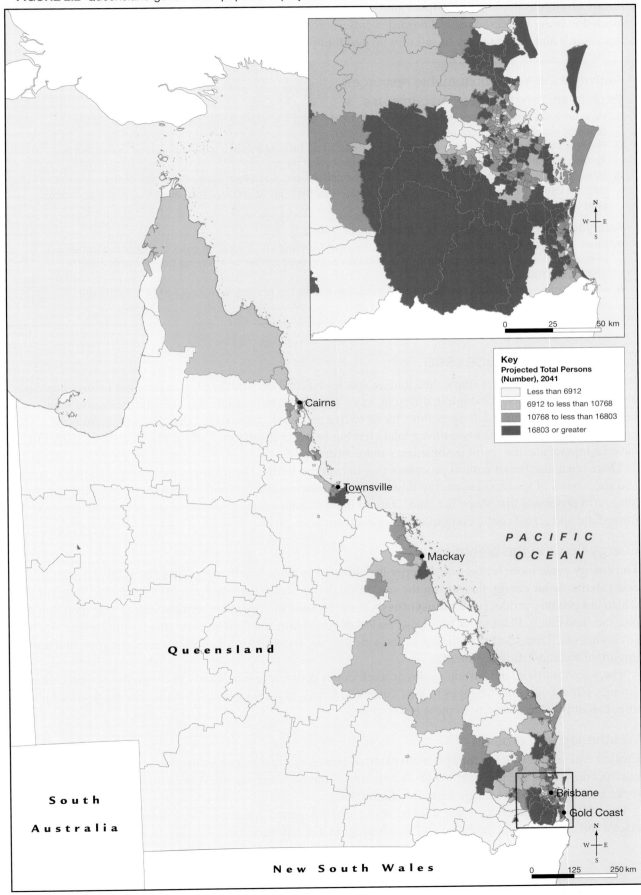

Source: © The State of Queensland Queensland Treasury 2019

Climate change and land cover change

Climate change is predicted to impact land cover significantly in coming decades. Climate change is expected to bring increased natural hazard events, increased and more unpredictable storm events, rising temperatures and rising sea levels, all of which will impact land cover (see chapter 1).

Activity 2.2a: Your consumable resources

Explain and analyse your resource use

1. How do you consume resources? Think about the different ways to categorise resources and list the different ways that you use them.
2. Research a typical item in your household such as a car, bike, television, mobile phone or computer, and itemise the different resources that are used to make that item.
3. How does the consumption associated with your chosen product affect land use? Think of the resources that are used to create it, where they are sourced, how they get to you and how all of that might impact on land use.

Analyse the data and apply your understanding

4. Identify the different components used to make your item and see if you can work out what resources are needed to create these components (e.g. tyres on a car need rubber or electrical components need precious metals).
5. Use an online tour creator tool to create a map that shows the primary location of all of the different components used in your product.

2.2.2 Natural processes

We have seen the immense impact the human species has had on land cover over time but natural processes also impact land cover, although in different ways. Generally, human impacts are immediate and long-lasting: widespread clear felling of trees for us to live and work, huge mines that extract ore from the ground for processing, or erosion from land clearing or water diversion. Natural processes take longer to have an impact and the result is sometimes more sustainable.

There are a number of natural processes that impact the Earth. The most obvious is through plant growth and succession of species (i.e. energy flows in the biosphere). Erosion and deposition of sediment are two long-term processes that shape the land. Natural hazards are examples of natural processes that can have immediate and significantly consequential impacts on land cover.

Energy flows in the biosphere

The energy cycle models how solar energy is converted to plant-based energy and distributed around **ecosystems**. Solar energy flows from the sun, through our ecosystems and initially transfers to green plants, which are primary producers. **Photosynthesis** converts solar energy into chemical energy that then flows into the food chain. Plant growth and succession of vegetation is a slow process that has major benefits for environments. Trees, herbs, bushes, grasses, vines, algae, ferns and mosses all cover landscapes, and are important for stabilising sediment.

The water, nutrient and carbon cycles at work in our ecosystems constantly cycle this matter around and through our ecosystems. Changes to land cover can have significant impacts to these energy flows which, in turn, can impact productivity in these ecosystems.

Weathering and erosion

Erosion and weathering are the two main external processes (that is, not related to the heat from the Earth's interior) that shape the physical world. Weathering is the physical, chemical or biological breakdown of rocks and minerals into smaller parts (see figure 2.4).

Physical weathering occurs when larger rocks are broken into smaller rocks. Frozen water expanding in cracks and crevices can cause physical weathering, as can wind and rain.

FIGURE 2.3 Energy flows in ecosystems

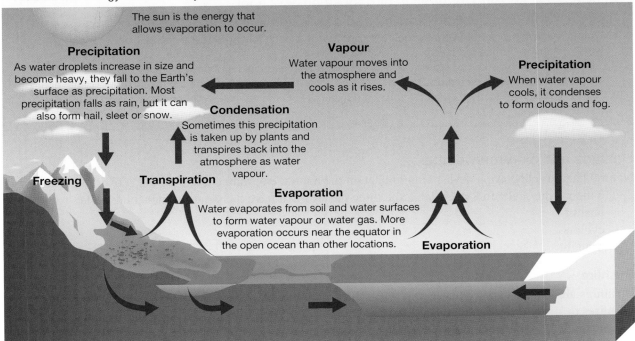

Chemical weathering occurs when rock is broken down due to chemical reactions. Carbon dioxide, oxygen and water often react with the atmosphere or minerals in the rock to cause this process.

Biological weathering occurs when living things break down rocks or minerals into smaller pieces. This can occur when tree roots destabilise and break down rock or when microbes release chemicals that slowly disintegrate rock. Although physical and chemical in process, these examples are driven by biological entities and are thus categorised as biological weathering.

Soil erosion is specifically the movement of soil, generally topsoil, from one part of the Earth to another driven by wind or water. Soil erosion can be natural or human-induced and can occur at small or large scales. Soil management is critical to the functioning of a society and entire civilisations have collapsed due to poor soil management practices.

FIGURE 2.4 Examples of physical, chemical and biological weathering

Deposition

Deposition of sediment occurs when eroded sediment is deposited due to energy loss. This usually occurs when the wind or water carrying the sediment slows down and drops the sediment. The inside bend of a river is an example of deposition of sediment. As the river moves around the bend the water moves faster on the outside but slower on the inside. This is why sediment gets deposited at river bends.

> **Activity 2.2b: Looking for erosion or deposition at school**
> **Research and communicate erosion and deposition in your local area**
> 1. Conduct a field trip around your school to try and find locations that show evidence of erosion or deposition, either natural or human-induced.
> 2. Select one location showing erosion or deposition that you observed and draw a figure to demonstrate how that erosion or deposition occurs. Ensure you use labels to fill out any information not indicated in the figure.
> 3. Research other types of erosion that you might expect in your local area. What are the different causes of that erosion?

Hazards and extreme weather

Natural hazards can have a devastating impact on the natural environment and on land cover in particular due to their large scale. In Queensland, the three hazards that have the greatest impact on the land are:
- bushfire
- cyclone
- drought.

Bushfire

Bushfires are a hugely important part of Australia's ecosystems, many having adapted to and relying on bushfires to stimulate growth. Some plant species in Australia have adapted to only release their seeds by heat and will experience massive regeneration after bushfire events. Bushfires considerably alter large swathes of land cover in very short periods of time, which means their impact on land cover can be enormous.

Australia identifies many bushfire events by name. Ash Wednesday (1987) and Black Saturday (2009) are two such examples. These events are named because they have a huge impact on the land and everything that inhabits it, including people. Their impact is felt everywhere on the continent. For Victoria and South Australia, summer and autumn are the high bushfire risk periods, while for New South Wales and Queensland late spring and early summer are the most dangerous. The Northern Territory and northern Western Australia experience most of their fire events in winter and spring. Geoscience Australia notes that the largest fires occur in remote parts of the country such as northern Western Australia, the Northern Territory and western Queensland, while the most damaging fires in terms of loss of life and economic impact occur on the edges of our cities and towns.

In late 2018, central Queensland experienced a particularly devastating bushfire season with more than half a million hectares of land burnt over a fortnight. An extended period of dry, hot weather, coupled with high temperatures, high winds and a delayed monsoonal season meant that the region was dangerously primed for a bushfire event. Nearly half a million hectares of land was burnt in November 2018 alone.

Cyclone

Queensland's cyclone season (November to April) usually brings between four and five cyclones into the Brisbane Tropical Cyclone Warning Centre's area of responsibility, which stretches from far north Queensland to just north of Newcastle in New South Wales. Although not all of the cyclones that enter this zone will cross the coastline, Queenslanders in high-risk areas are taught to be well-prepared for these events, and local and state government initiatives have been designed to ensure people's safety. However, the natural world cannot prepare for these events and when a cyclone does cross the coast it usually has devastating impacts on the land post-event.

Cyclones bring wind gusts of between 100 km/h and 280 km/h in extremely violent events. Winds moving at these speeds can cut through huge swathes of the land and destroy almost everything in their path. Human developments are often blown to pieces in strong cyclone events and forests, crop lands and other vegetation can be stripped from their roots and sediment. Floods associated with cyclonic rainfall can destroy everything in their path, as seen in Townsville and the far north of Queensland's floods of early 2019 (see figure 2.5).

FIGURE 2.5 North-west Queensland in 2019, (a) before and (b) after flooding

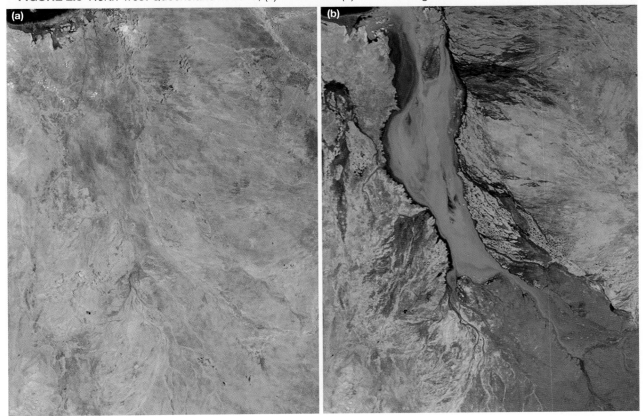

Source: NASA

Drought

Drought events aren't as instantaneous as other hazard events and the processes that lead to them occur over long periods of time. Two main processes lead to drought: prolonged periods of reduced **precipitation** in an area and increased evaporation. Drought can dry up the soil, which prevents or inhibits plant growth. This land cover change is also felt by animals and people, because there is less vegetation and crops for eating.

2.2.3 The world around us

Some of the prominent types of land cover in Queensland include urban areas, coasts, rivers, and forests and grasslands. These are all different land covers, therefore they are shaped by different processes.

Urban areas

Urban areas, by their very existence, are examples of land cover changes. Development, new infrastructure and increased residential living all require the removal of land cover to be replaced with another. Increasing urbanisation will only increase this sort of land cover change and the impacts that go with it.

Coasts

Coastal landforms are affected by erosion and deposition. Erosion can bring sediment from the tops of mountains many kilometres inland to river mouths along the coast. Longshore drift moves sediment from river mouths to beaches and along coastlines. Wind then moves this sediment up the beach, forming dune systems behind beaches.

Dunes systems are most impacted by deposition of sediment from the ocean, driven by onshore winds. Sediment moves up the beach via wind energy and is then deposited as the wind carrying it slows due to low-lying grasses, vegetation up the beach or human features such as fences, showers or even a beach towel. Gradually, this deposited sediment accumulates and grows into a small sand dune, known as a primary dune. Further dunes can develop behind this dune and they are known as secondary and tertiary dunes (see figure 2.6). Generally, vegetation gets denser further away from the beach.

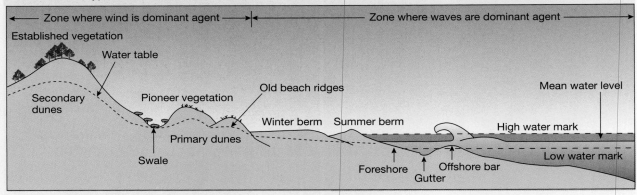

FIGURE 2.6 Typical dune structure

Rivers

Rivers and waterways form an important part of the **water cycle** because they drive through the landscape, taking water downstream from headlands at higher ground. When enough water collects in a catchment (the area that feeds a river system), gravity can turn it into a trickle by pulling it down towards sea level. As these trickles become tributaries and move downstream, more water collects into streams or rivers and over time all of this water can accumulate to create huge flows of water like the mighty river systems of the world, including the Nile, the Amazon, the Yangtze, and even the Murray River or the Brisbane River. Erosion of the Earth can be seen in all river systems as the movement of water has carved its way through stone and soil over many years. This erosion leads to unique river valley shapes such as U- and V-shaped valleys in the upper reaches and meanders, oxbow lakes, and deltas in the lower reaches of river systems. Deposition can be seen in the inner corner of river bends as the water slows and deposits sediment.

Forests and grasslands

In any part of the earth where vegetation covers the ground, the flow of energy from the sun, water and nutrients drive growth. Forests are areas with enough precipitation to support a wide variety of plant and animal species. The main types of forest biome are tropical, temperate and boreal (see chapter 1).

FIGURE 2.7 A typical Queensland forest

Grasslands, also known as **savanna**, **prairie**, **steppes** or **tundra**, are areas of the Earth that get enough rainfall to support grasses but not enough to support more advanced plant species. Grasslands are defined by their limited precipitation and by the dominance of grass species in the absence of larger plant species. Occasionally, fire can pass through grasslands, causing widespread land cover change but also renewal of some of the vegetation in these areas as the burnt grass is fed back into the soil, making it rich and fertile.

FIGURE 2.8 A small plot in the Amazon basin

Forests and grasslands are filled with vegetation that converts solar energy through photosynthesis. Primary and secondary producers consume that energy and decomposers break down the producers when they die. All of these elements cycle energy through our ecosystems.

Deserts

Deserts are areas that experience little precipitation and have little vegetation or ground cover. They are not completely barren and can occur in tropical, temperate and polar regions. Deserts make up almost 30 per cent of the Earth's land surface, with most of the world's deserts located 30 degrees north or south of the equator. There are no deserts on the equator itself, where you might expect them to exist due to the heat and abundant sunlight.

At the equator, the sun's energy warms up water, which causes it to evaporate and be carried into the air. Large, warm masses of moist air are driven towards the poles from the equator by the **Coriolis effect** and these cells slowly deposit their moisture in the form of rain while they start to cool and sink. This sinking tends to occur around 30 degrees north and south of the equator. On the ground, these areas experience relatively cool, dry sinking air, which leads to the formation of deserts as precipitation is greatly reduced at these areas. The poles both share similar environmental conditions that promote deserts, although they are significantly colder (see figure 2.9).

Deserts are the harshest environments on Earth and are noted for their unique and creative plant and animal adaptations. These adaptations allow these species to survive in a low-precipitation and low-vegetation environment.

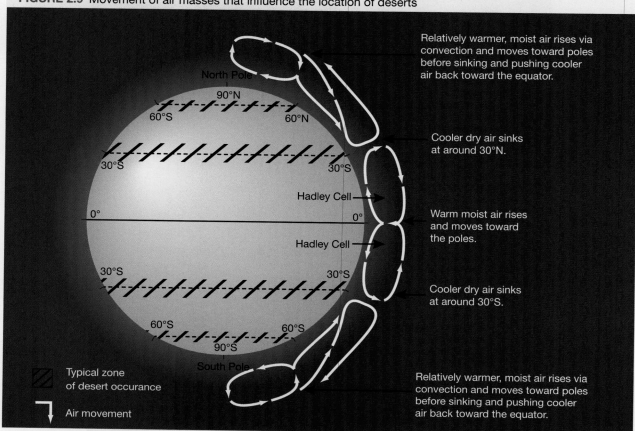

FIGURE 2.9 Movement of air masses that influence the location of deserts

2.3 The spatial pattern of land cover change

As discussed, land cover can change for a wide variety of reasons, both natural and human. Where these changes occur depends on a range of factors. Topography, climate and economic reasons based on human needs and wants drive most land cover change.

2.3.1 The spatial pattern of agriculture

Agriculture refers to the cultivation of plants and animals for use by humans. We use agricultural products every day. For instance, most of what we eat is sourced from crop production or by slaughtering domesticated animals, and most products, including leather, natural clothing fibres, medicines and oils, are created through agricultural production.

The Queensland Farmers' Federation defines agriculture as, '... Any activity connected with the growing of food, fibre, timber and foliage including, but not limited to, cropping, intensive horticulture, animal husbandry, intensive animal industry, animal keeping, **aquaculture**, permanent plantation, wholesale nursery, production nursery, roadside stall, winery and rural industry; and also including ancillary activities concerned with accommodation of farm workers, visitors and tourists; the storage of water; irrigation and drainage works; the storage of equipment for the production and transport of agricultural products; and the on-farm processing, packaging, storage and sale of agricultural products.'

It was agriculture, specifically the cultivation of wild grains and the domestication of livestock, that progressed humans from the hunter-gatherer lifestyle to a more sedentary one. This occurred somewhere between 10 000 and 12 000 years ago at the end of the **Pleistocene Ice Age** and the start of the **Holocene period**. Subsistence farming allowed the farmer to feed their family and to swap or sell any surplus product that they had. Markets were places where rural communities could buy and sell produce and other items, and they were the precursor to today's 'farmer's market', where many different types of produce, foods and craft are available in one location.

It is agriculture that has had the largest global impact on land cover change. Ever since humans embraced agriculture, we have required land to grow crops, for somewhere to live, to store supplies and produce, to transform the raw materials into something worth selling and to graze livestock. Modern, large-scale, industrial farming businesses operate over huge areas of land and produce most of our food, crops and animal products, while having a significant impact on the natural environment and rural communities (see figure 2.10).

FIGURE 2.10 Agricultural crop yields globally, 2014

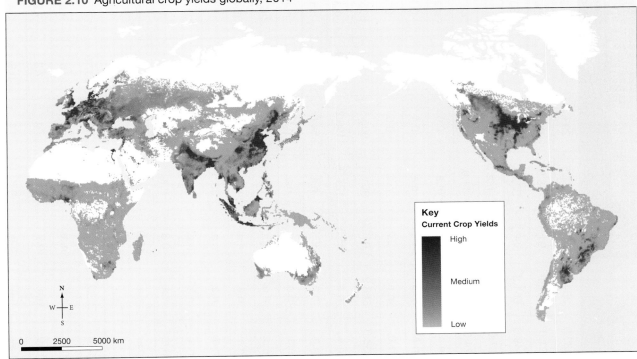

Source: Max Roser and Hannah Ritchie. Our World in Data.

Queensland has a 60 000 year history of agriculture, which began with the Indigenous Australians. The myth that Indigenous Australians were a predominately hunter-gatherer society has left a great lack of knowledge about Australia's ancient history of agriculture. Indigenous Australians ploughed fields and harvested crops, mainly tubers such as yams, though not in the regimented way recognised by Europeans when they arrived in Australia. Indigenous Australians also carefully selected and preserved foods and irrigated crops, and there is even evidence that they transported fish stocks and harvested aquatic animals like eels with sophisticated aquaculture infrastructure such as those found at Budj Bim in Victoria.

In Queensland's agricultural industry, what is grown now depends on geography and climate. Different crops grow in different areas and different animals are suitable for large-scale production in different areas.

Queensland is dominated by three main crops: wheat, sugar cane and sorghum. In 2016–17, there were 622 000, 385 000 and 250 000 ha of land devoted to each, respectively. As shown in table 2.1 following, the amount of land set aside to grow different crops can change over the years due to farmers adapting to different climatic conditions, fluctuations in prices for crops, resource availability (particularly water availability), changing consumer needs and desires, and changing technology.

TABLE 2.1 Area by crop, Queensland, 2006–7 to 2016–17

Crop		2006–7	2007–8	2008–9	2009–10	2010–11	2011–12	2012–13	2013–14	2014–15	2015–16	2016–17
						('000 ha)						
Sugar cane		383.0	354.5	375.7	370.5	293.4	326.1	316.5	359.1	360.1	361.5	385.7
Cereals for grain:	Barley	81.0	112.9	91.9	68.5	94.0	80.0	89.7	105.8	125.5	137.7	149.0
	Sorghum	449.0	661.2	538.2	333.4	434.6	435.5	430.6	355.6	546.7	363.1	249.9
	Maize	27.0	n.a.	49.3	n.a.	37.0	42.9	40.2	26.2	30.7	30.1	35.3
	Oats	24.0	20.3	7.3	11.6	13.6	17.9	48.0	23.3	65.3	36.2	47.4
	Millet	n.a.	n.a.	n.a.	n.a.	0.0	n.a.	n.a.	n.a.	n.a.	n.a.	n.a.
	Rice	n.a.	n.a.	0.0	n.a.	0.4	0.3	0.1	n.a.	2.2	3.9	0.3
	Triticale	1.0	n.a.	0.2	n.a.	0.5	1.1	0.1	n.a.	2.2	3.9	0.3
	Wheat	638.0	668.9	1,019.9	961.7	905.5	953.2	865.9	758.0	633.5	611.1	622.2
Legumes for grain:	Chickpeas	n.a.	n.a.	85.5	127.9	199.2	148.9	n.a.	n.a.	n.a.	251.6	n.a.
	Mung beans	n.a.	n.a.	36.2	n.a.	54.6	n.a.	n.a.	n.a.	n.a.	101.5	n.a.
	Navy beans	n.a.	n.a.	n.a.	n.a.	n.a.	n.a.	n.a.	n.a.	n.a.	n.a.	n.a.
Crops and pastures for hay:	Pure lucerne	n.a.	n.a.	n.a.	30.5	n.a.	25.9	20.3	19.9	19.2	16.9	12.1
	Other pastures	n.a.	n.a.	n.a.	n.a.	n.a.	n.a.	n.a.	32.7	32.7	28.7	22.1
	Cereals	n.a.	n.a.	n.a.	n.a.	21.5	12.1	17.5	26.1	35.0	40.0	27.5
	Other crops	n.a.	n.a.	n.a.	19.1	18.7	19.8	22.8	22.8	30.7	20.0	12.7
Oilseeds:	Safflower	n.a.	n.a.	0.8	n.a.	1.1	n.a.	n.a.	n.a.	n.a.	n.a.	n.a.
	Soybeans	n.a.	n.a.	18.9	n.a.	3.9	n.a.	n.a.	n.a.	n.a.	n.a.	n.a.
	Sunflower	n.a.	n.a.	17.4	n.a.	10.2	n.a.	n.a.	n.a.	n.a.	n.a.	n.a.
Cotton		44.0	28.7	78.4	87.6	258.6	237.6	170.3	140.0	73.2	93.8	203.1
Linseed		n.a.	n.a.	n.a.	n.a.	n.a.	n.a.	n.a.	n.a.	n.a.	n.a.	n.a.
Peanuts (in shell)		12.0	n.a.	9.9	n.a.	7.1	7.3	7.2	5.5	4.7	4.8	4.6
Tobacco		n.a.	n.a.	n.a.	n.a.	n.a.	n.a.	n.a.	n.a.	n.a.	n.a.	n.a.

n.a. = not available

Source: © The State of Queensland Queensland Treasury 2019 based on ABS data

Because of Queensland's climate and geography, the state is uniquely placed to grow sugar cane, which is used in everything from raw sugar through to ethanol supplements for petrol. Ethanol fuel releases fewer **greenhouse gases** than regular unleaded petrol, but when agriculture and production costs are factored in the gains are somewhat less than promoted.

Activity 2.3a: The impact of cattle grazing

Explain the impacts on land cover

1. Research the impact that cattle grazing has on Queensland. Consider the economic value of the industry, its employment potential and the environmental impact of land clearing and grazing itself.

Present and analyse the data

2. Use an online mapping tool from the Queensland Government such as QLUMP to create and examine where cattle grazing and cropping occurs in Queensland.
3. Describe where cattle grazing and cropping occur across Queensland using appropriate terminology. Represent some of your data using a chart, graph or map. Consider the best method of data representation for your needs.
4. What natural processes are potentially interrupted by cattle grazing in Queensland?
5. What potential methods could be applied to reduce the negative impacts of cattle grazing in Queensland? Conduct research online to find methods in use in other countries.

Activity 2.3b: Graphing queensland's crops

Explain and synthesise crop data

Go to the website of the state government department responsible for agriculture to answer the following questions. (At time of print this was the Department of Agriculture and Fisheries.)

1. What percentage of Queensland is used to grow cereal crops?
2. What are the top three crops in terms of production in the most recent year or season you can find?
3. Create a line graph showing the yield of cotton in Queensland over time. Ensure your graph adheres to common chart conventions.
4. Suggest reasons for the fluctuations in production in some crops over time.

Research and analyse impacts on the environment

5. Research the specific impacts of cotton growth on the environment and respond to the following statement, either for or against, using evidence to support your argument.

   Queensland should ban cotton farming across the state until more efficient methods of growing it can be found.

2.3.2 The spatial pattern of urbanisation

Urbanisation occurs when when people move to urban areas often from rural areas, and when the proportion of a population living in urban areas increases. Globally, 55 per cent of people live in a city and this proportion has been steadily increasing over the last century. The World Bank predicts that by 2045, 6 billion people will live in cities. Cities in Africa and South America are predicted to have the highest rates of urbanisation in the near future according to the UN.

In Australia, the trend over the past 100 years has been one of increasing urbanisation. There are five urban areas in Australia with populations of more than one million: Sydney, Melbourne, Brisbane, Perth and Adelaide.

A local government area (LGA) is a portion of land, including all the people and buildings on it, that a local government is responsible for. Queensland's top ten largest growing LGAs to the end June 2017 are shown in figure 2.11. The top seven LGAs in terms of population increase were Brisbane, Gold Coast, Moreton Bay, Sunshine Coast, Ipswich, Logan and Redland, all of which are located in the south-east corner of the state. The other three LGAs are all in the regional locations of Cairns, Toowoomba and Townsville. Although more people were added to Brisbane's population because it is a larger city, Ipswich's growth rate in 2017 of 3.2 per cent far outstrips Brisbane, which registered only 1.9 per cent growth over the year.

FIGURE 2.11 Top ten largest growth LGAs in Queensland, to 30 June 2017

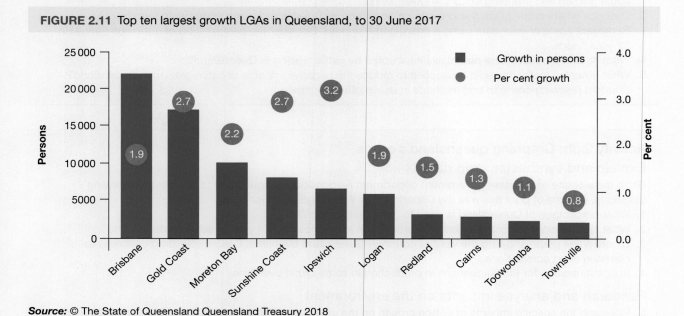

Source: © The State of Queensland Queensland Treasury 2018

Urbanisation occurs at the edge of cities and towns but can also occur inside these places. We see urbanisation where farmland, forests and grasslands are removed for housing developments on the urban fringe. These areas also see infrastructure development as roads, highways and rail lines are built to connect us to other places. Inside cities and towns, we see infill development, where population densities are increased to accommodate more people in the same area. This is usually done by replacing houses with townhouses or apartments, or replacing existing apartments with even bigger ones.

Figure 2.12 shows the scale of urbanisation in south-east Queensland over ten years. In 1999, Brisbane, the Gold Coast and the Sunshine Coast were distinct urban entities but in 2019 these cities started to merge, particularly along the Pacific Motorway as people and businesses took advantage of the convenience of a major transport route running through all three cities.

FIGURE 2.12 Satellite images of south-east Queensland, (a) 1999 and (b) 2019

Source: © State of Queensland 2018.

Resources

Video eLessons SkillBuilder: Understanding satellite images (eles-1643)

SkillBuilder: Interpreting satellite images to show change over time (eles-1733)

Interactivities SkillBuilder: Understanding satellite images (int-3139)

SkillBuilder: Interpreting satellite images to show change over time (int-3351)

2.3.3 The spatial pattern of deforestation

Forests cover around 24 per cent of Earth's landmass and they can be found where a moderate amount of precipitation falls; that is, enough precipitation to support grasses and stabilisers, undergrowth, trees and a richer amount of plant life than grasslands or savanna. Forests can be classified as either tropical, temperate or boreal (polar), depending on the temperature at each location. **Deforestation** occurs when forests are removed, most often for development or to use the wood as a resource. The creation of forests can take hundreds of years, but deforestation occurs immediately and is therefore semi-permanent. It is dramatically increasing in modern times because the global population has increased so rapidly in the past century. From 2001 to 2017 there was a total of 337 million hectares (MH) of tree cover lost globally.

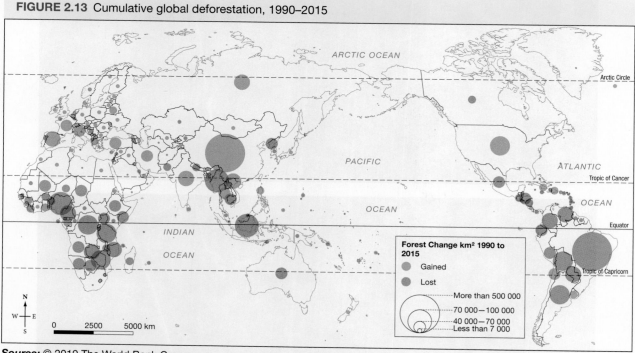

FIGURE 2.13 Cumulative global deforestation, 1990–2015

Source: © 2019 The World Bank Group

The areas that had the greatest area lost to deforestation from 1990–2015 were Brazil, Indonesia, Nigeria and Myanmar. China was leading the way in forest growth during that period, but growth was also seen in the US, India, Russia and Spain (see figure 2.13).

Australia is in the midst of a deforestation boom period, particularly in Queensland (see figure 2.14). In 2018, Queensland cleared more land than every other state in Australia combined. According to the World Wide Fund for Nature (WWF), Australia is on track to lose 3 MH of land in the 15 years between 2018 and 2032.

Queensland has an exceptionally high rate of land clearing. As a state, Queensland has exceeded the rate of tree clearing seen in the Brazilian Amazon. In 2018, the WWF placed Queensland alongside regions like Borneo and the Congo Basin as hotspots of vegetation clearing. At its deforestation peak in the mid-2010s, Queensland was clearing the equivalent of more than 1000 football pitches of woody vegetation per day. More than 392 000 ha were cleared per year across the whole state in 2017 and 2018.

FIGURE 2.14 Spatial distribution of woody vegetation clearing in Queensland, 2017–18

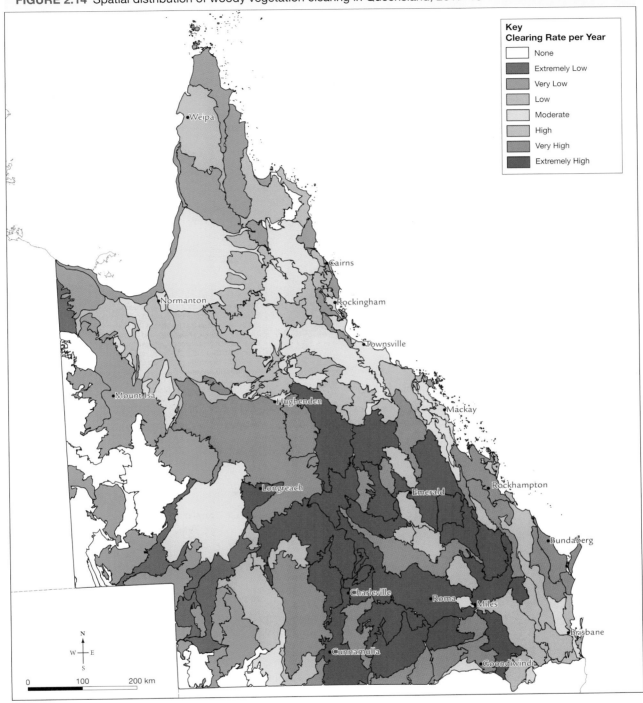

Source: © State of Queensland, 2018

Land clearing in Queensland predominantly occurs in the Brigalow Belt and the Mulga bioregions of central southern Queensland. In 2017–18, per year, 204 000 ha were cleared in the Brigalow Belt and 106 000 ha were cleared in Mulga country. These regions contain most of the catchment area for Queensland's section of the Murray–Darling Basin and clearing in these areas can have significant impacts, not just in Queensland, but also further downstream in the catchment. Of equal importance was the 47 per cent of woody vegetation cleared in Queensland in 2016–17, which came from catchment areas along the east coast that feed the Great Barrier Reef (see figure 2.15).

FIGURE 2.15 Clearing rates across (a) the Great Barrier Reef catchment area, (b) by bioregion and (c) by drainage division in Queensland, 2016–18

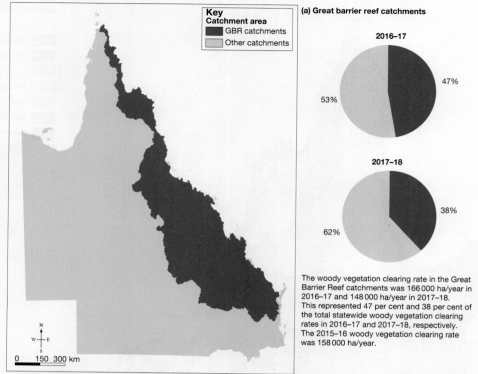

The woody vegetation clearing rate in the Great Barrier Reef catchments was 166 000 ha/year in 2016–17 and 148 000 ha/year in 2017–18. This represented 47 per cent and 38 per cent of the total statewide woody vegetation clearing rates in 2016–17 and 2017–18, respectively. The 2015–16 woody vegetation clearing rate was 158 000 ha/year.

Source: © State of Queensland, 2018

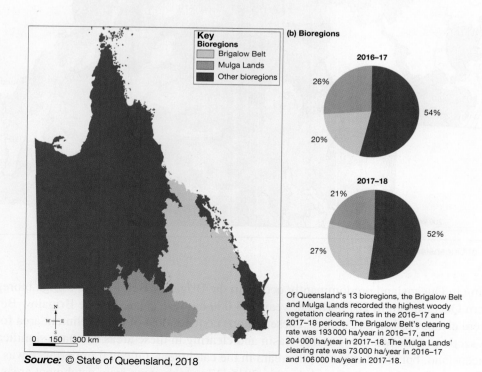

Of Queensland's 13 bioregions, the Brigalow Belt and Mulga Lands recorded the highest woody vegetation clearing rates in the 2016–17 and 2017–18 periods. The Brigalow Belt's clearing rate was 193 000 ha/year in 2016–17, and 204 000 ha/year in 2017–18. The Mulga Lands' clearing rate was 73 000 ha/year in 2016–17 and 106 000 ha/year in 2017–18.

Source: © State of Queensland, 2018

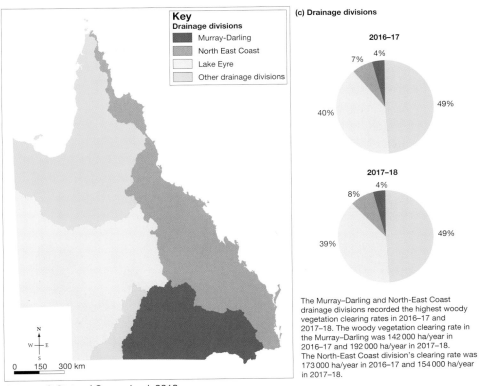

Source: © State of Queensland, 2018

2.3.4 The spatial pattern of mining

Mining is the removal of materials, sediments and ores from the ground in order to process them. It has long played a role in Queensland's history, from Indigenous Australians' use of stone and other mineral resources to modern, large-scale mining operations that cover many hectares and extract millions of tonnes of material from the ground. Mining is a key driver in Queensland's economy and shapes the identity of many locations across the state.

There are mining operations all over the state of Queensland, and the location of the mines depends on the resource being mined. Weipa in far north Queensland is one of the world's largest bauxite producers. Coal is found in abundance in central Queensland and the Darling Downs.

2.3.5 The spatial pattern of hazards

Due to their size, Australia and Queensland experience more natural hazard events than most places on Earth. The three hazards that impact us the most are drought, bushfires and cyclones, and all occur because of our location and climate.

Drought

Australian has one of the lowest average rainfall measures on Earth, so drought is a hazard that many Australians are familiar with. Drought is a hazard that can impact everywhere in Queensland, but particularly inland regions where rainfall is less certain. The 24-month rainfall deficiency map in figure 2.16 shows that central southern Queensland and south-west Queensland have experienced some of the lowest rainfall totals in recorded history. In 2019, the hottest five years on record globally were the previous five years between 2014 and 2019. This means that drought is a hazard that will continue to influence Queensland and one to which we must adapt.

FIGURE 2.16 24-month rainfall deficiency for Australia, 2017–19

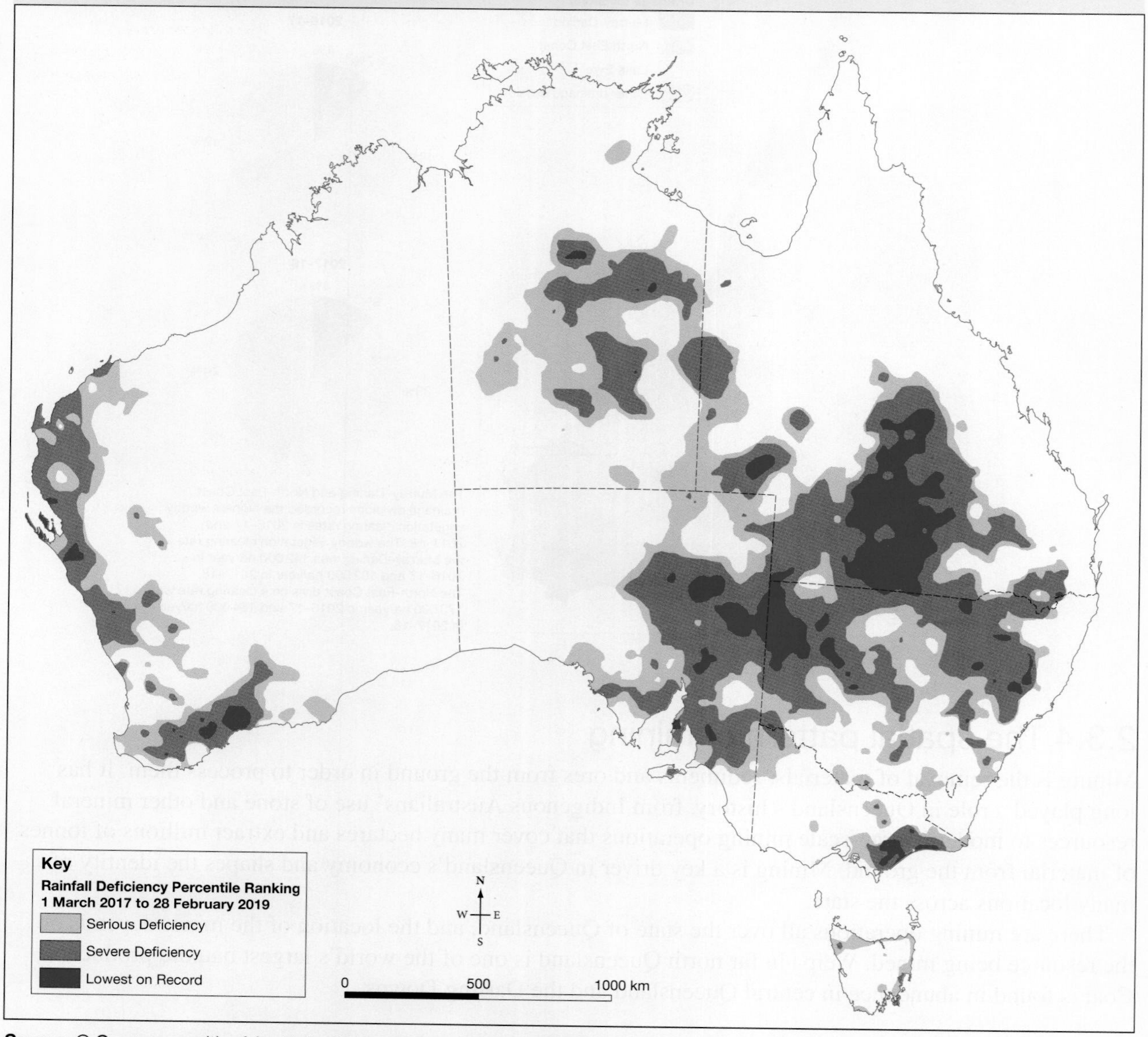

Key
**Rainfall Deficiency Percentile Ranking
1 March 2017 to 28 February 2019**
- Serious Deficiency
- Severe Deficiency
- Lowest on Record

Source: © Commonwealth of Australia, 2019, Australian Bureau of Meteorology

Bushfire

Like drought, bushfire events are frequent in Queensland due to climatic conditions and the availability of fuel. Arid areas in the west of the state and grasslands covering the central areas are less prone to bushfire events but areas of woody vegetation, particularly in the north, are more likely to experience bushfires. However, due to the major population centres being on the east coast, particularly in the south-east, these bushfire-prone areas may not always be the highest risk in terms of damage and loss of life. The areas of rural–urban convergence, where the population increases, tend to see more death, injury and destruction of property.

FIGURE 2.17 Queensland's bushfire prone areas

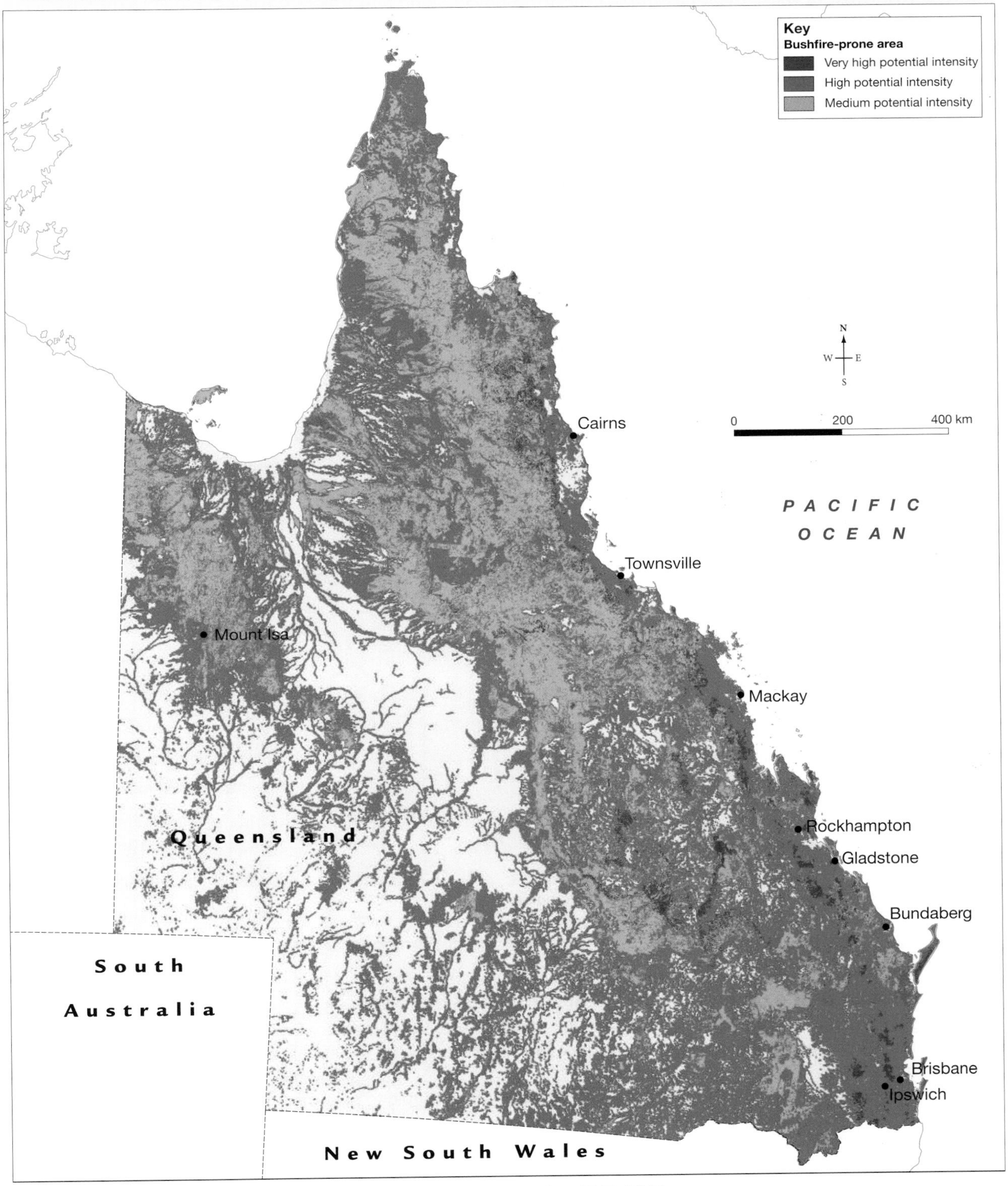

Key
Bushfire-prone area
- Very high potential intensity
- High potential intensity
- Medium potential intensity

Source: © Commonwealth Scientific and Industrial Research Organisation, 2015–2017

Cyclone

Tropical cyclones are as much a part of Queensland's summers as the beach and the sun. On average, four to five cyclones cross the Queensland coast every cyclone season between November and April, and they bring high winds, huge amounts of rainfall and, potentially, storm surges and post-cyclone flooding (see figure 2.18). Most of these cyclones impact far north Queensland and it is rare for a cyclone to directly influence the southern part of the state, although it has happened in the past, most famously when Cyclone Wanda brought extensive flooding to south-east Queensland in 1974.

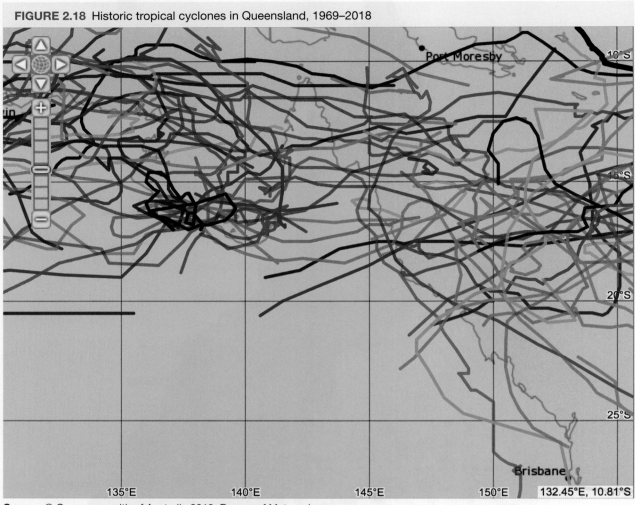

FIGURE 2.18 Historic tropical cyclones in Queensland, 1969–2018

Source: © Commonwealth of Australia 2019, Bureau of Meteorology

Activity 2.3c: Creating a thematic map
Create a map and explain natural hazard and land cover change

1. Using an online mapping tool like the Queensland Globe, Google My Maps or ScribbleMaps, create a simple thematic map showing the varying levels of risk for the different types of natural hazards referred to in subtopic 2.3.
2. Add data for one more natural hazard that isn't listed in the subtopic to your map.
3. Describe in a paragraph which areas of Queensland have the highest overall risk for land cover change due to natural hazard.
4. In another paragraph describe which areas have the lowest risk for natural hazard induced land cover change.

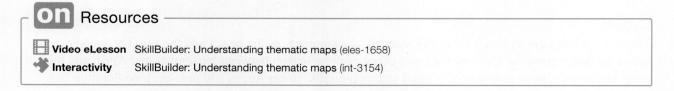

Resources

- **Video eLesson** SkillBuilder: Understanding thematic maps (eles-1658)
- **Interactivity** SkillBuilder: Understanding thematic maps (int-3154)

2.4 The effects of land cover change

Land cover changes affect different environments in different ways and these changes impact on people as well. Physical changes to land cover often profoundly change how that lands fits into the ecosystem and its ability to cycle energy, carbon, water and nutrients.

Changes also often affect how people perceive and use the land. Obvious changes include the building of a housing development on formerly agricultural land, clearing forest for a road or a cyclone decimating vegetation.

This subtopic will examine different impacts of land cover changes in different environments. These examples are by no

FIGURE 2.19 Land cover in urban areas

means an exhaustive list of the effects that one might see in an environment as it is changed, naturally or anthropogenically or both. These examples should be a starting point for you; your fieldwork will then allow you to see and experience the impacts of land cover change in real life.

2.4.1 The impacts of land cover change in urban areas

Urban areas will impact on the land differently depending on how they are developed. Where agricultural land is removed for housing, the impact is enormous, but where a house is replaced with an apartment block, the impact is not as significant.

Land cover change in urban areas can be as obvious as the removal of native vegetation to build an apartment block, road, housing estate, new shopping centre or sports stadium. In these cases, where vegetation is removed there are significant impacts on energy, water, carbon and nutrient flows as well as the loss of habitat for native fauna.

Land cover changes in urban areas can also be subtle as cities expand and change according to their size, economy and the technology that is available. In a social sense, significant land cover change in urban areas can affect people's sense of community and the aesthetic appeal of an area. Residents often see these being reduced by change, especially by large-scale forces like **gentrification**, which see the **demographics** and **socioeconomic status** of an area undergo massive changes over only a few years. Sometimes development provides improved services for people in urban areas, such as more parks, and better facilities and services, like transport connections. Areas of increased urbanisation also create centres of economic, social, educational and cultural prosperity.

Changes to urban areas are almost always justified on the basis that they will increase economic activity in an area. More residents means more people shopping, doing business and paying tax in that jurisdiction, which means more money for the local, state or federal government overseeing the area.

The economic downside is the expensive cost of development, roads and other infrastructure. Population density is lower in outer suburban areas, due to lower population and larger physical area, and so it costs significantly more to provide infrastructure such as power lines, sewage, garbage collection, etc, to these areas compared to providing infrastructure in more densely populated areas, such as the city centre.

Socially, it is the equity of urban and suburban living that is the biggest issue. Access to resources, services and facilities can vary between different areas and it could be political representation or socioeconomic status that is the greatest influencer in how these resources are allocated.

Activity 2.4a: Analyse your local area
Using online tools to analyse data
Use the **Queensland Globe online** interactive tool in the Resources tab to complete the following activity.
1. Go to your local area and evaluate its access to the following:
   - Transport connections
   - Health facilities
   - Educational facilities
   - Recreational facilities
   - Commercial facilities
2. Compare your local area with a classmate's to see the similarities and differences in access to services. Can you explain any differences that you see?
3. Make three recommendations that would help to improve your local area based on your findings and at least one secondary data source, such as the census.

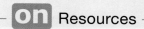 Resources

 Weblink Queensland Globe online

2.4.2 Effects of change on our coasts

The coast is an important place for many Australians, and is considered a place of recreation and beauty. Approximately 80 per cent of the population lives within 50 km of the coastline. Our coasts are delicate ecosystems that need to be in balance and changes to our coastline can adversely affect those ecosystems.

Environmental impacts of changes to our coasts

Development on coasts removes vegetation and habitat, and can damage or wipe out dune systems. Coastal vegetation is important but vulnerable, and acts as a stabiliser, and its removal can lead to erosion. Dunes are components of the sediment budget on low-energy coasts and their removal can have widespread impacts in other areas of the ecosystem.

Change to our coastlines can interrupt natural geographic processes including longshore drift. The training walls at the mouth of the Tweed River in New South Wales were built to protect the commercial shipping operations that used the river by keeping the mouth clear. Unfortunately the development of the training walls led to a significant reduction of sediment along the beaches of the southern end of the Gold Coast as the process of longshore drift was artificially interrupted.

The lack of sediment on the beach led to the removal of huge amounts of sand, the destruction of many kilometres of dunes and the exposure of roads and houses that were built close to the beaches. The loss of sediment was also exacerbated by a series of storms and cyclones that hit the region in the early 1970s. Efforts to address these problems are discussed in section 2.6.2.

2.4.3 Effects of change on our rivers and waterways

Waterways play a vital role in moving water across the land, especially on dry **continents** like Australia. Land cover changes can have profound impacts on our rivers and waterways, which can in turn impact agricultural and rural communities along the waterways.

CASE STUDY — THE MURRAY–DARLING BASIN

The Murray–Darling Basin covers 1 061 469 km² across five states or territories: Queensland, New South Wales, the ACT, Victoria and South Australia. Queensland alone accounts for approximately one quarter of its total area. The Basin drains into the Murray and Darling Rivers. The Darling River has its headwaters in southern Queensland and northern New South Wales, and the Murray River begins in the New South Wales and Victorian Alps. The Murray and Darling Rivers bring water from the Great Dividing Range west and then south, where the river system empties into the ocean in Goolwa, in south-eastern South Australia. The total water flow in the Murray–Darling Basin has averaged 24 000 gL per year, although this can fluctuate depending on rainfall.

There are two areas where humans have had the biggest impact on the Murray–Darling Basin. One is the diversion and overuse of water for agricultural purposes, and the other is the removal of vegetation along the rivers' **riparian zone** and in the Basin itself.

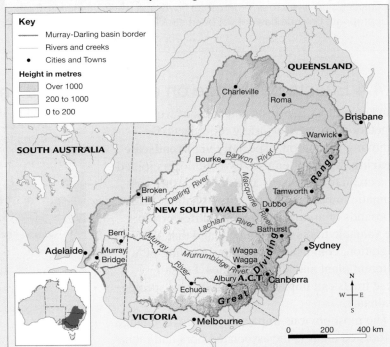

FIGURE 2.20 The Murray–Darling Basin

Source: © Commonwealth of Australia Geoscience Australia 2018. Redrawn by Spatial Vision.

The overuse of water for agricultural purposes removes water from the water cycle and from ecosystems along the waterway. This has led to many different impacts including loss of water volume at different periods in the Basin's history and at different locations, **algal blooms** all along the system, increased pollution and lowering of water quality because the system cannot flush itself with the regularity it requires. Land cover change from inside the Murray–Darling Basin can lead to increased runoff, which can carry sediment, nutrients and chemicals like pesticides into the water system.

Riparian zones act as a buffer that can slow down runoff and keep some of the heavier materials, such as sediment or pesticides, out of the waterway. Removal of these zones removes the buffer and can increase the amount of damaging materials that flow into the water system.

In early 2019, millions of fish along stretches of the Murray–Darling system died and floated to the surface in multiple, spontaneous events. This demonstrates the connectivity between different elements of the river system.

This event occurred because land clearing led to an increase in nutrients in the waterway and overuse of water reduced the total amount of water. The situation was exacerbated by an extended drought across south-eastern Australia, and then an algal bloom and rapid temperature change extracted all of the oxygen from the waterway which led to the rapid death of millions of fish. Water extraction, drought and vegetation loss all played a part in the final outcome.

Potential solutions to the changes

Since 1915, various arrangements have been in place that have outlined how the Murray–Darling's water resources will be used and managed. In the early 1990s, a shift towards a market-based approach allocated water resources to users and water use was then capped. As water quality and volume have continued to fall along the Murray–Darling, greater

FIGURE 2.21 The Menindee fish deaths in early 2019

Source: GRAEME MCCRABB / AAP

consideration has been given to the value of water flows for environmental uses in the Basin and these moves have likely been exacerbated by the continual closing and shifting of the mouth of the Murray–Darling. Other, more radical, solutions call for the reregulation of water allocations along the Murray–Darling, with greater consideration given to environmental flows and uses. Ultimately, there is strong evidence to show that an appropriate balance has not yet been struck between the environment and economic users of the system's water.

2.4.4 Effects of change on our forests and vegetation

Land cover loss in forests and areas of woody vegetation can have a variety of impacts. By removing vegetation, especially via **broadscale clearing**, which often uses two bulldozers joined by a chain to clear large areas of land, the most dramatic impact can be the loss of habitat for thousands of animal species. Wildlife corridors are valuable tracts of vegetated land that allow migration, colonisation and interbreeding of different plant and animal species. Loss of these tracts can impact **biodiversity** and species resilience.

FIGURE 2.22 Broadscale clearing chain, usually connected to two bulldozers

Source: © 2019 Green Collar Group

The most obvious impact of land cover change in forested areas is the associated flora and fauna species loss. Australia is a significant place in terms of species **diversity** on the planet. Twelve per cent of the world's vertebrate species are found in Australia, of which four out of five are native, and we have experienced the known loss of at least 90 species extinctions since European colonisation. University studies have found that habitat loss affects 74 per cent of Australia's threatened species; land cover changes for agriculture and urbanisation constitute most of that loss. The loss of native species increases the opportunities for species to be introduced into an ecosystem, which can further damage the environment.

Removal of vegetation can lead to an increase in erosion. Vegetation stabilises loose soil and its removal can leave that soil exposed. Exposed soil will be removed by wind or water so erosion is almost guaranteed.

As water rolls across the land it picks up everything in its path. Exposed soil is collected and moved by the water, and gravity, into a waterway that eventually feeds into a river system. This leads to increased sediment in the waterway, which reduces the waterway's ability to convert solar energy into oxygen via algae. This increased runoff of soil usually brings any nutrients that may be used in the catchment area. Pesticides and herbicides add considerable nutrients and other chemicals to the waterway as they are collected by water and moved from the land into the waterways.

Streambank erosion is the wearing away of the banks of a waterway, and is a huge problem in Queensland that affects the Great Barrier Reef. The Queensland government is tackling the problem by using satellite imagery to identify problem streambanks and implementing a program that could include:
- physical remediation (including piloting of different techniques)
- mulching, revegetation and fencing
- native grass seed production
- grazing land management improvements
- traditional owner engagement and training
- scientific research and monitoring to understand the nature of gullies
- communication with stakeholders
- encouraging employment and tourism opportunities.

Activity 2.4b: Spatial change in land cover
Explain and analyse change in land cover using online tools
Use the online interactive tool **QImagery** in the Resources tab to complete the following activity.
1. Select an area of Queensland and use QImagery to view historical aerial imagery in that area. Use Surfers Paradise on the Gold Coast to see immense change if your area hasn't changed much.
2. Represent the spatial changes to land cover at the local level using simple spatial technologies like the Queensland Globe, Google My Maps or ScribbleMaps.
3. Identify the implications for environments and people of the changing land cover, including on spiritual and cultural features of value for Aboriginal peoples and Torres Strait Islander peoples.

---

on Resources

Video eLesson SkillBuilder: Comparing aerial photographs to investigate spatial change over time (eles-1750)
Interactivity SkillBuilder: Comparing aerial photographs to investigate spatial change over time (int-3368)
Weblink QImagery

2.5 Connections between people and physical systems

The connection between land cover, place and local people is strong, important and necessary. For many millions of years, humans and our ancestors have been the most influential species on the planet and our impact on land cover at a local scale has been critical to our survival as well as our progress and development as a species. As a result humans have a natural connection to the land — it shapes us as much as we shape it.

2.5.1 Aboriginal and Torres Strait Islander connections to the land

The land is of great significance to Indigenous Australians. It is a place of birth and death, it sustains, it teaches, it helps to explain the world and beyond, it defines spirituality, it is all ancestors, and it changes and grows as people do. People's relationship with the land is reciprocal — one both takes and gives back.

Aboriginal and Torres Strait Islander peoples' understanding of features and elements of their local areas grows from their connection to place. Aboriginal people have used the land as a central theme in the Dreaming as a means of explaining and understanding how the world was created and why many geographic environments, processes and features exist.

This connection to the land can be physically expressed through:
- the natural environment
- Dreaming sites
- sacred sites
- archaeological sites (for instance quarries, middens, skeletal remains, etc)
- ceremonial sites
- water holes
- burial grounds.

Observing a natural process over time, such as the life cycle of a dragonfly, and teaching about that aspect of the environment allowed Indigenous people to learn about everything that naturally surrounded that process. These included the weather, geography, water and water flows, vegetation, flowers flowering, and other animals in the food chain that relied on or coexisted with the dragonfly. All of this knowledge originates from the land. Seasonal calendars are one way some of this knowledge can be represented visually. These calendars change across different parts of the country, but an example from far north Queensland is shown in figure 2.23.

FIGURE 2.23 Seasonal calendar for the Yirrganydji people of far north Queensland

Source: © Yirrganydji Community 2019, hosted by Bureau of Meteorology, http://www.bom.gov.au/iwk/calendars/yirrganydji.shtml

The Lost Girl

The Aboriginal connection to place can be represented by the traditional story, *The Lost Girl*. It tells the story of a girl who wanders away from her family, falling asleep in the shade of a rock. When she wakes, she is alone and does not know how to get back to her camp. She calls for help but is not heard. The girl tries to find her way back, finding sustenance on her way. She takes water from the river when thirsty and berries when she is hungry. The sheltered rocks keep her warm and a crow flying in the night sky then guides her back to camp. Her family rejoice on her return, and when asked if she had been scared, she responds

> 'How could I be frightened? I was with my Mother. When I was thirsty, she gave me water; when I was hungry, she fed me; when I was cold, she warmed me. And when I was lost, she showed me the way home.'

Activity 2.5a: Consider the evidence

Comprehend and explain the importance of land

1. Read the story of *The Lost Girl*.
2. List the ways the land is useful to the girl on her adventure.

Research artefacts and your local area for evidence of connection to land

3. Consider some of the artefacts that are on display (physically if possible as well as online) in the University of Queensland's Anthropology Museum. What about them points to a more intimate connection to the land in Indigenous Australian communities than most non-Indigenous communities? Consider what the artefacts are made from, their uses and design.
4. Research your local area and see if you can identify some traditional sources of food or other resources from local flora and fauna.

> **Resources**
>
> 🔗 **Weblink** University of Queensland's Anthropology Museum

Agriculture

For more than 60 000 years Indigenous Australians have undertaken clever and sustainable land management practices that have been based on an intimate and deep connection to the local area and the land beyond. Aboriginal and Torres Strait Islander people managed the land using age-old techniques that have been tested and retested under Australian conditions. In regards to agriculture these techniques include:

- harvesting and cultivating crops to use as a source of food and a commodity for trade. Vast tracts of what is commonly thought of as useless land were once covered in fields of Australian grains that are slowly being reintroduced to growers.
- damming of waterways strategically for water storage and agricultural purposes
- threshing, harvesting, drying and milling of grains for cooking (flour) occurred in numerous locations around Australia as evidenced by records from European explorers
- transportation of seed stocks across the landmass in order to increase genetic diversity of plant species
- deliberate selection of seed stocks to create varieties that are tailored to their local environmental characteristics, soils, climate, rainfall and even topography
- storage of surplus grains for trade and use in non-seasonal periods
- capture and storage of fish, eels and other marine life through drainage manipulation and the construction of fish traps, widely regarded to be the oldest human-made structures in existence.

All of these innovative and ingenious methods of managing the land were not recognised by Australia's colonial settlers as they sought to make Australia as British as possible. They used British agricultural methods instead, which although they did not realise it, were entirely unsuitable for Australia's climate and soils. However, people are now seeing the benefits of the traditional Indigenous methods and they are starting to be used again.

Housing and development

As discussed, Indigenous Australians were not exclusively a hunter-gatherer society as is often portrayed. There is evidence from across the continent that shows Aboriginal and Torres Strait Islander peoples developed permanent settlements for housing and storage, even in areas that are now considered entirely inhospitable. Indigenous housing was sophisticated, using timber frames, thick thatched roofing and ingenious design features to make the most of the local climate and conditions. All of these features were documented by European explorers and surveyors. On one of his earlier expeditions to what is now known as the Darling River, Charles Sturt observed and described a settlement with at least 70 huts capable of holding at least 15 people each. He wrote, '[The houses] were made of strong boughs fixed in a circle in the ground, so as to meet in a common centre; on these there was … a thick seam of grass and leaves and over this a compact coating of clay. They were from eight to ten feet [3–4 m] in diameter, and about four and a half feet [1.5 m] high, the opening into them not being larger than to allow a man to creep in. These huts also faced north-west and each one had a smaller one attached.'

FIGURE 2.24 Archaeological dig in Western Australia

Source: Peter Veth

Fire

Indigenous Australians have used fire in a variety of ways, including to enable access to parts of the environment, for hunting and for ceremonial purposes. It was also used as a land management tool — fires were deliberately started in cooler months to reduce fuel loads, stimulate habitat growth for some animal species and to allow fire-sensitive plant communities to thrive. Studies have shown that effective use of fire to undertake continuous but small-scale burning can actually increase biodiversity in an area. These methods create a mosaic of burnt and unburnt areas across the landscape, which is in contrast to one-off large scale burns that may be undertaken by non-Indigenous authorities. These fires wipe out large swathes of the land, tend to reduce biodiversity and increase stress on the environment.

> **Activity 2.5b: Indigenous fire management**
> **Research and communicate an argument**
> 1. Research and list Indigenous fire management strategies.
> 2. Argue whether or not you think these strategies should be employed in managing fire in modern Australia. Use evidence to support your claim.

2.6 How can we mitigate our negative impacts on the land?

Since European colonisation, many methods of land management have been implemented in Australia with varying levels of success. As mentioned, early European land management techniques and land use strategies, particularly around agriculture, were almost directly imported from Britain but these methods often proved ineffective in Australia, with its different heat, rain, soil and other unique features. In fact, European techniques were often hostile to the Australian landscape, flora and fauna.

Over time, these methods have been refined, tested and evaluated and we are slowly developing better ways to manage our land use and how we change land cover in the Australian context. This subtopic will explore some of these methods in the different ecosystems covered in this topic, but these strategies represent only some of the ways that we can manage our impact on the land.

2.6.1 Urban areas

Increased regulations around how we develop and use the land have delivered the greatest benefit to urban development and land cover change. When Europeans began to establish settlements across Australia, many of which are now large cities and towns, development in these areas was not as regulated as we see now. Building styles and inclusions were not mandated, environmental considerations were not taken into account and developers took advantage of this for hundreds of years. It isn't hard to find examples in most Australian towns and cities of developments that are still standing that are unsustainable, unsuitable and possibly even damaging to the local area.

By effectively undertaking and applying research on sustainable development and urbanisation, most cities and towns have adopted regulations that reduce

FIGURE 2.25 South Bank in Brisbane, Queensland, is an example of effective urban design

Source: © State of Queensland 2018.

negative impacts on the environment, society and the economy. This isn't true of all government decision-making but the changes in pollution control and environmental and air quality in our cities and towns over time shows that regulation is effective. Some of these regulations include:
- environmental controls at building sites
- land clearing laws
- requirements for development applications for development projects
- environmental impact considerations on developments
- sustainability considerations in government procurement
- changes to automobile and fuel technology.

Changes to the way individuals and communities treat the world can also have considerable impacts on sustainability in an area. Methods of land management that attempt to replicate natural processes are becoming more popular as their effectiveness is demonstrated. Permaculture is a fantastic example of how we can manage local environments in such a way as to maintain and enhance the environment, society and the economy while replicating natural processes.

Permaculture

Permaculture is a set of design principles that recognises the interconnectedness of ecosystems. It is a systems approach to agriculture that attempts to replicate natural systems in order to achieve sustainability. The concept has spread beyond agriculture to include culture more broadly. The principles of permaculture can be applied in fields such as agriculture, systems ecology, environmental design, town planning, urban design, and water and soil resource management.

FIGURE 2.26 Permaculture in action

Permaculture principles are being adopted widely these days. They are applicable at scales from the individual, to the local area, to the national. There are twelve design principles of permaculture that allow users to considerably alter their impact on the environment.

1. Observe and interact: Understand nature's patterns in order to design appropriate solutions to problems.
2. Catch and store energy: Develop systems to capture and store energy in all parts of development.
3. Obtain a yield: Grow for a purpose.
4. Apply self-regulation and accept feedback: Improve your work and outputs.
5. Use and value renewable resources and services: Decrease dependence on non-renewable resources.
6. Produce no waste: Use resources efficiently to eradicate waste.
7. Design from patterns to details: Observe and replicate natural patterns where possible.
8. Integrate rather than segregate: By putting the right things in the right place, relationships develop between those things and they work together to support each other.
9. Use small and slow solutions: Smaller solutions encourage greater attention to detail in an area or project.
10. Use and value diversity: Diversity reduces vulnerability to threats and enhances the unique nature of the environment in which it resides.
11. Use edges and value the marginal: The interface between things is where the most interesting events take place. These can be the most productive elements in the system.
12. Creatively use and respond to change: Observation and understanding of local environmental patterns allow us to adapt to change quickly.

Activity 2.6: Understanding permaculture

Explain and analyse information

1. What is permaculture?
2. Research some of the different applications of permaculture in cities and list three that you think are achievable in your school, household or community.
3. What needs to change in order for your ideas from question 2 to be accepted widely?

Communicate your action plan

4. Take action! Write a letter and present your proposal to someone in a position of authority. Remember, if you put it in writing, they are more likely to respond.

2.6.2 Coasts

Managing coastal development has its own unique challenges. One of the best examples of effective management of issues related to change along coasts is the Tweed River Sand Bypass System (TRSBS). It is considered innovative because of the way in which it solves a problem by replicating a natural process.

The problem relates to the building of the Tweed River training walls in order to maintain a deep channel for commercial fishing boats. The training walls interrupted the process of longshore drift. This meant beaches along the southern Gold Coast lost considerable amounts of sand as the sediment was collected on the southern side of the Tweed River mouth. The natural process of longshore drift would usually move this sediment in a northerly direction along Australia's east coast.

After a series of storms swept through the Gold Coast area in the early 1970s, massive erosion occurred on these exposed beaches, causing damage to the dunes, property and infrastructure. Giant boulders and car bodies were imported to the beach to try to reduce the sand loss. This widespread and shocking destruction of an iconic Australian beach signalled the beginning of intent by the public to solve the problem. Nearly thirty years later, the TRSBS was installed and operational.

FIGURE 2.27 Tweed River training walls

Source: Skyepics

The TRSBS is a sand transport system that collects sand from the southern side of the river and then pumps it to the north side, by going under the river. Once the sand is moved, the waves and currents move sand naturally to the Gold Coast beaches. Almost immediately after installation, sand levels along the southern Gold Coast beaches were stabilised and sand budgets along these beaches returned to normal levels. The training walls are still in place, and fishing and boating vessels can enter the Tweed River unimpeded (see figure 2.27).

2.6.3 Rivers

Management of river systems has been ineffective in the past when it failed to recognise the interconnectedness of ecosystems and the importance of management techniques in parts of catchments away from the waterway and its riparian zone. Greater emphasis on how farmers manage their lands in recent times has yielded significant improvements to water quality, although there are still many improvements to be made, as demonstrated by the problems in the Murray–Darling Basin.

Rejuvenation of river ecosystems can happen through a range of methods. In general, rebuilding of the riparian zone and reinstating water flow, where possible, are the two most important actions to positively manage rivers.

2.6.4 Forests and vegetated areas

Managing the impacts on vegetated areas has probably been one of our nation's greatest failings since European colonisation and with the rate of land development only increasing, it will be one of the hardest areas to manage. Prior to European colonisation, over 30 per cent of Australia's landmass was covered in forest, but today that figure lies at around 16 per cent. Most vegetation loss has been to accommodate grazing and crop lands. Although in recent years most Australian states have reduced tree clearing rates, Queensland's increase in tree clearing for pasture has overshadowed the improvements made in other states.

Activism can encourage action and awareness of issues related to land cover change in forested or vegetated areas. Because environmental outputs in supply chains (i.e. pollution) are not factored into the economics of a product, action at a corporate level on environmental issues can come with pressure from the public. Consumer driven actions such as organised boycotts or awareness campaigns can force businesses to carefully consider their actions and those of their suppliers. This can eventually force changes in corporate behaviour. One example of this is the recent campaigns about palm oil. The production of palm oil requires the clearing of pristine rainforest vegetation across the globe to create palm oil plantations. Activists have generated awareness of these issues and this has allowed consumers to choose products without palm oil to take action against the environmental damage done by palm oil plantations. Many of the world's major manufacturers and food and beverage makers have reconsidered where they source the oils used in their products and this has reduced demand for palm oil. It is hoped this will lead to a slowing of the land clearing that was happening at alarming rates, particularly across south-east Asia.

FIGURE 2.28 Australia's forest extent, 2013

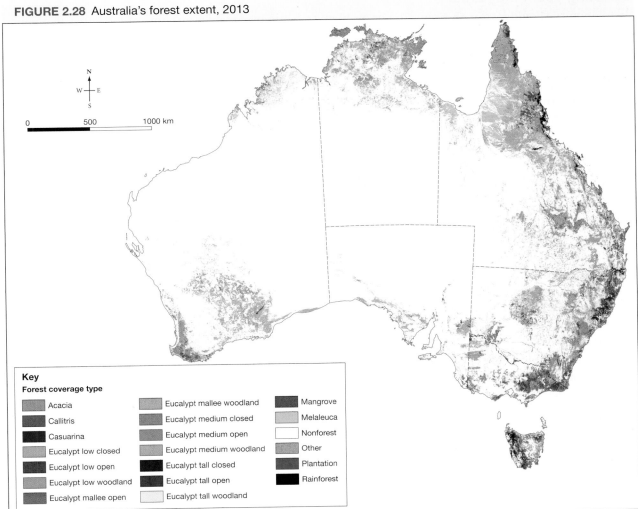

Source: Australian Bureau of Agricultural and Resource Economics and Sciences, Forests of Australia 2013

2.7 Fieldwork in the local area

2.7.1 Introduction to fieldwork

Fieldwork is an important part of geography and offers the best of what the subject is about: getting out to experience and explore the real world. Fieldwork is always built around an investigation and the data you collect in the field should give you a better understanding of the geographical issues at hand.

The geographic inquiry model is a framework used by geographers to help structure an inquiry. It is a great way to help you organise your thinking as you undertake planning an investigation and it can be used to address issues from local to global scales.

FIGURE 2.29 Geographic inquiry approach

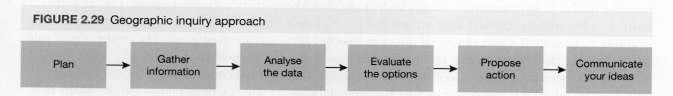

2.7.2 Plan

Selecting your study area

Every inquiry should start with a question or problem so begin by considering some of the issues that you have been studying in class. Try and frame your question in such a way as to allow you scope to collect data, analyse it, make conclusions and respond with meaningful proposals. This might mean adjusting your geographic area or the scope of your problem.

The geographic area you select is important. If you select an area that is too large you may give yourself too much to cover in your data collection and your analysis can also get unwieldy. Having too much to cover also means that you may not be able to complete your analysis and develop proposals within the word limit.

Planning and structure

Use the geographic inquiry model to structure your investigation. Develop key questions and start with as many focus questions that you can think of — these will expand and change as you start to collect data and research, and you get a better understanding of the issue in question.

Develop your key and focus questions to help you to organise your research and data collection. These key questions usually take the form of:
- *What* and *where* is the issue or problem at hand?
- *How* and *why* does it occur?
- *What* are the environmental, social and economic *impacts*?
- *What* are the proposals and *solutions*?

The key questions can then be narrowed down through the use of focus questions that concentrate your research into smaller areas. These focus questions can change and should be updated as you gather more information through your research and fieldwork.

The example below outlines how you might structure an inquiry into a new housing development in an area on the urban fringe. As you begin to research your focus questions you will find more information and uncover more questions to ask, which means your investigation should be reviewed and will be constantly evolving.

- What and where is the development?
- How and why is the development going to influence the local landscape?
- What are the impacts of the development on the local area, economy and society?
- What solutions can be put in place to ensure that the development has minimal impact on the local landscape?

Now is a good time to begin thinking about what sort of data you will need to collect and what sort of research you will do. Will your **primary data** answer all of your questions? What types of **secondary data** should you collect and where will you need to go to get it? Think about the organisations that might host this data.

Activity 2.7a: Planning your fieldwork
1. Decide on which type of land cover you will be working on in your fieldwork report.
2. Come up with an issue or problem that you want to address for your fieldwork study. Try and frame it as a question (e.g. Should Brisbane have a second airport runway that removes mangroves? How can we address the spread of the cane toad in western Queensland?).
3. Create a rough outline of your investigation using the geographic inquiry model key questions.
4. Underneath each key question generate some focus questions to get your investigation started. Write down where you might find the answer to each focus question. Is it primary data or secondary data?
5. What data collection methods will you have to employ to undertake your fieldwork?

Format options

There are many ways that you could present your report and you should think about the method that will best communicate your information. The traditional, typed and printed hardcopy report will be the most common format, but if you have the skills you might like to consider using an alternative format, such as the ones listed below. It is important to remember that whatever format you use, you should always ensure that you fulfill the requirements of the report and that you are able to include all relevant information. For instance, a tweet thread may not give you enough scope to include all relevant information but a website would allow you to elaborate and demonstrate your knowledge and skill.

Possible format options include:
- hardcopy report
- story map
- video
- game
- interactive web presentation
- Prezi or PowerPoint presentation.

2.7.3 Gather information

In undertaking fieldwork, you should endeavour to collect as much relevant data as possible to help your investigation. Primary data is collected first-hand. In research, primary data holds the most weight because it is data collected by the author to help them answer specific research questions — it is data collected for that purpose. Secondary data is collected by someone else and can include census data, government data and information, and research findings. Your analysis, and subsequently your decisions, solutions and proposals, will be evidence-based provided you have collected appropriate primary data and sourced reliable and relevant secondary data.

One of the hardest points in your fieldwork will be choosing what to measure so that you can give yourself the best opportunity to show off your analysis skills. Sections 2.7.4 to 2.7.7 suggest fieldwork ideas for urban areas, coasts, waterways and vegetated areas.

2.7.4 Fieldwork ideas for urban areas

Urban areas are areas of constant change that most people have some experience with. We see shops change hands, apartment blocks go up, parks become housing estates and roads get upgraded. Consider the following activities to collect useful primary data.
- Urban transects
- Land use maps
- Survey
  of services (e.g. transport options or health services)
- Survey
  of resident opinions on an issue or proposed solution
- Field sketching
- Analysis of satellite imagery over time
- Gentrification and urban consolidation
- Public space survey

FIGURE 2.30 Sharing data collected in the field

2.7.5 Fieldwork ideas for coastal areas

Coasts come in high and low energy versions. You may collect data in one or both depending on the issue being investigated. The following activities give you opportunities to collect meaningful data while on the coast.

Low energy coasts:
- Dune transects
- Dune profile
- Land use mapping
- Field sketching
- Wave analysis
- Wind analysis
- Sediment analysis
- Investigating beach protection measures (e.g. groynes, sand-bypass, etc).

High energy coasts:
- Field sketching
- Wave analysis
- Wind analysis.

A sample investigation Coastal Fieldwork Booklet is provided in the Resources tab.

2.7.6 Fieldwork ideas for rivers or other waterways

Waterways offer many opportunities to measure and collect useful field data. Everything from an intermittent stream to a raging river can give geographers great data. Try the following examples.
- Flow analysis — velocity
- Water quality
- Land use mapping
- Historic imagery comparison
- Stream bed measurements or survey
- Field sketching

FIGURE 2.31 Water quality data collection

2.7.7 Fieldwork ideas for vegetated areas

Most schools have access to some sort of vegetated area and a properly remote patch of forest may not be too far away. The following fieldwork ideas should be achievable for most.
- Canopy analysis
- Quadrat analysis
- Vegetation mapping
- Infrastructure mapping
- Field sketching
- Soil sampling.

See the SkillBuilders in the Resources tab for practical guides on various geographic techniques.

> **Activity 2.7b: Collect your data**
> 1. Collect your field data.
> 2. You will need to represent your data visually. Think about what the most suitable method will be to represent your data. Consider what message you want to convey to and how you can do that most efficiently without confusion.
> 3. Find some online tools to help you represent and analyse your data. There are many simple online data tools that you can use to help your reader understand your issue.

2.7.8 Analyse the data

Once you have collected your data you need to analyse it. There are a number of methods to do this. Generally, you are looking for patterns in the data that are related to the topic at hand. Sometimes these patterns are self-evident (i.e. poor water quality near an industrial area) but sometimes additional data needs to be collected or analysed in conjunction with your data to draw out the message.

The primary data that you collect and secondary data that you select and interpret should be able to help you explain the geographic processes at play in your issue and why they have led to the source of any geographical challenges. Your analysis should also focus on the different impacts that could occur as a result of your challenge continuing. You should be able to extrapolate from your analysis to generalise about the impacts of your issue.

Think about different ways to visualise your data so that you can demonstrate your understanding of the processes involved in each issue and the analysis that you have undertaken to reach your findings. You can use traditional methods to present data such as maps, tables, charts, field sketches or photographs, but you can also consider other methods, such as data visualisations or digitally generated maps using spatial technologies.

Spatial technologies

Spatial technologies are the tools of geographers. They help us answer 'where' questions and can help us illuminate some of the issues that geographers study. They are used in a range of geographical and non-geographical ways, including environmental management, species tracking, weather and climate analysis, oceanic and inland water monitoring, economics and history.

Geospatial technologies help us to measure, analyse and represent spatial or geographic information, usually digitally.

In regards to your field study, you could use spatial tools to:
- examine land use and/or land cover from a satellite image
- investigate public transport and cycling infrastructure in the area around your place of study
- visualise census and other demographic data using Google Maps or the ABS online mapping tools
- visualise environmental data like earthquakes or species extent using online digital tools
- map data you have collected in the field, such as water quality data or survey results.

FIGURE 2.32 Geospatial technologies in the field

Source: © State of Queensland 2018.

2.7.9 Evaluate the options

Evaluating is the art of selecting or choosing between options. You should consider a range of proposals and then choose the best one. It is important that you have clear and measurable reasons as to why you have chosen the option you chose. This may come down to the data you have collected while in the field.

2.7.10 Propose action

The ultimate reason you are undertaking all of this fieldwork and collecting all of this data is to make recommendations in order to effect change around the issue you have selected. Consider developing a set of criteria that you can use to help you evaluate your proposals.

2.7.11 Communicate your ideas

The last task is to put all of your planning, research, data, analysis and decision-making into a coherent piece of writing that can be presented. You will be given specific guidelines on how to structure and present your findings so make sure you understand what is required of you. A sample and scaffolded field report are provided in the Resources tab.

Before you submit your work, take time to review it or have someone you know read it to ensure it makes sense and flows logically. Having someone who hasn't seen it before can be advantageous as they will be looking at it with a neutral point of view, while you are very familiar with your work. Good luck!

Activity 2.7c: Write a field report
1. Write your field report.
   Two useful resources are provided in the Resources tab of your eBook. These are:
   - A scaffolded Field Report Booklet
   - A sample Coastal Fieldwork Booklet.

Resources

Digital documents
Sample investigation: Scaffolded Field Report Booklet (doc-31437)
Sample investigation: Coastal Fieldwork Booklet (doc-31436)

Video eLessons
SkillBuilder: Constructing and describing a transect on a topographic map (eles-1727)
SkillBuilder: Constructing a land use map (eles-1755)
SkillBuilder: How to develop a structured and ethical approach to research (eles-1759)
SkillBuilder: Creating a survey (eles-1764)
SkillBuilder: Using advanced survey techniques — interviews (eles-1742)
SkillBuilder: Understanding satellite images (eles-1643)
SkillBuilder: Interpreting satellite images to show change over time (eles-1733)
SkillBuilder: Constructing a basic sketch map (eles-1661)
SkillBuilder: Constructing a field sketch (eles-1650)

Interactivities
SkillBuilder: Constructing and describing a transect on a topographic map (int-3345)
SkillBuilder: Constructing a land use map (int-3373)
SkillBuilder: How to develop a structured and ethical approach to research (int-3377)
SkillBuilder: Creating a survey (int-3382)
SkillBuilder: Using advanced survey techniques — interviews (int-3360)
SkillBuilder: Understanding satellite images (int-3139)
SkillBuilder: Interpreting satellite images to show change over time (int-3351)
SkillBuilder: Constructing a basic sketch map (int-3157)
SkillBuilder: Constructing a field sketch (int-3146)

2.8 Review

2.8.1 Chapter summary

This chapter has covered some key points about responding to local land cover transformations.

Processes resulting in land cover change
- Two anthropogenic processes that affect land cover are urbanisation and resource exploitation.
- Natural processes impact land cover, and erosion and deposition of sediment are two long-term processes that shape the land around us.
- Coasts have a number of different physical processes at play that affect coastal landforms such as erosion and deposition, dune systems, rivers and waterways, and longshore drift.

The spatial pattern of land cover change
- Topography, climate and economic reasons (based on human needs and wants) drive most land cover change through agriculture, which has the largest impact on land cover change.
- Australia has a high proportion of the population living in cities and the trend over the past 100 years and into the future will be increasing urbanisation, which affects land cover such as forests and farms as they give way to houses, industry and commercial land use.
- Deforestation occurs when forests are removed, most often for agricultural development, urbanisation or to use the wood as a resource.
- Queensland has an exceptionally high rate of land clearing, predominantly occurring in the Brigalow Belt and the Mulga regions of central-southern Queensland.
- Mining has long played a role in Queensland's history and still contributes to changes in land cover.
- Natural hazards can have a devastating impact on the natural environment and on land cover in particular due to their scale. In Queensland the most common natural hazards are bushfire, cyclone and drought.

The effects of land cover change
- Land cover changes affect different environments in different ways and these changes also impact on people.
- Development on coasts removes vegetation and habitat, and causes erosion and damage to dunes.
- Land cover changes can have profound impacts on our waterways and rivers, leading to impacts on agricultural and rural communities along waterways.

Connections between people and physical systems
- Impacts of land cover change in forested areas include the associated flora and fauna species loss, and erosion.
- The connection between land cover, place and local people is strong, important and necessary.
- The land is of great significance to Indigenous Australians. Their understanding of features and elements of local areas grows from their connection to place.
- Indigenous Australians have an ancient history of agriculture including harvesting and cultivating crops for food, and damming of waterways for water storage and agricultural use.
- Contrary to previous portrayals, Indigenous housing was often permanent and existed across Australia.
- Indigenous Australians have used fire for land management, to enable access to parts of the environment, for hunting and for ceremonial purposes.

How can we mitigate our negative impacts on the land?
- Ways to mitigate our negative effects on the land include undertaking and applying research on sustainable development and urbanisation, permaculture, and proactively managing coasts and rivers.
- Activism is a method that can encourage action and awareness of issues related to land cover change in forested or vegetated areas.

Fieldwork in the local area
- Fieldwork is an important part of geography and the geographic inquiry model can help provide a framework for investigation.

2.8.2 Key questions revisited
You should now be able to answer the following questions.
- What is land cover change?
- How is land cover important to us environmentally, socially and culturally?
- What processes are involved in local land cover changes?
- What are the impacts of local land cover changes?
- How can we manage local land cover changes to maximise sustainability?

UNIT 4
MANAGING POPULATION CHANGE

What will the world look like in 30 years? Both globally and in Australia, demographic challenges impact how and where we live. Population changes constantly over time and space, across the world. Some population change represents people on the move, but whether this move is voluntarily or involuntary, both affect where people come from and where they are going.

In this unit you will study the impact of population change and look closely at rural, remote or urban places in Australia through geographic inquiry, and study the challenges faced there.

CHAPTER 3 Population challenges in Australia (Unit 4, Topic 1) 119

CHAPTER 4 Global population change (Unit 4, Topic 2) 185

3 Population challenges in Australia

3.1 Overview

3.1.1 Introduction

In this topic, you will analyse the geographical processes that have resulted in population change in Australia over time and space. This analysis will focus on a range of demographic indicators, including birth and death rates, fertility rates, life expectancy and migration patterns.

You will be involved in the recognition of spatial patterns of population change in Australia and of the implications of these changes for people and places over time. As part of your study, you will investigate a particular demographic challenge for a place in Australia through geographic inquiry, using **primary data**.

As a result of this investigation, you will be able to understand the factors that influence demographic change in places in Australia and the challenges that arise. You will be asked to propose actions to manage these challenges. Through your geographic inquiry, you will be able to understand the nature of population changes over time and the impact on the resulting needs and **resources** of a community.

FIGURE 3.1 Australia's population is growing and changing

3.1.2 Key questions
- What is demography? What are the key demographic concepts?
- What are Australia's population patterns and trends?
- What demographic processes are responsible for changes in population in Australia?
- What factors influence changes in population in Australia, both spatially and over time?
- What are the implications of population change for people and places in Australia?
- What are the demographic characteristics of a particular place in Australia?
- What challenges occur because of demographic change in a particular place in Australia?
- What actions might be taken to manage these challenges?

3.2 Demographic concepts

3.2.1 What is demography?

Demography is the study of population, especially human population. It involves statistical analysis of such characteristics as **population size** and **composition, distribution** across space and the processes through which populations change over time. Figure 3.2 provides information on the relative population size of the world's countries using a cartogram (a map with diagrammatic statistical information, in this case the population of the world's countries drawn according to their population size rather than their geographic area). As you can see, the world's most populous countries are China and India. Australia, in comparison, is one of the world's smaller countries in terms of population size.

FIGURE 3.2 World population, 2018

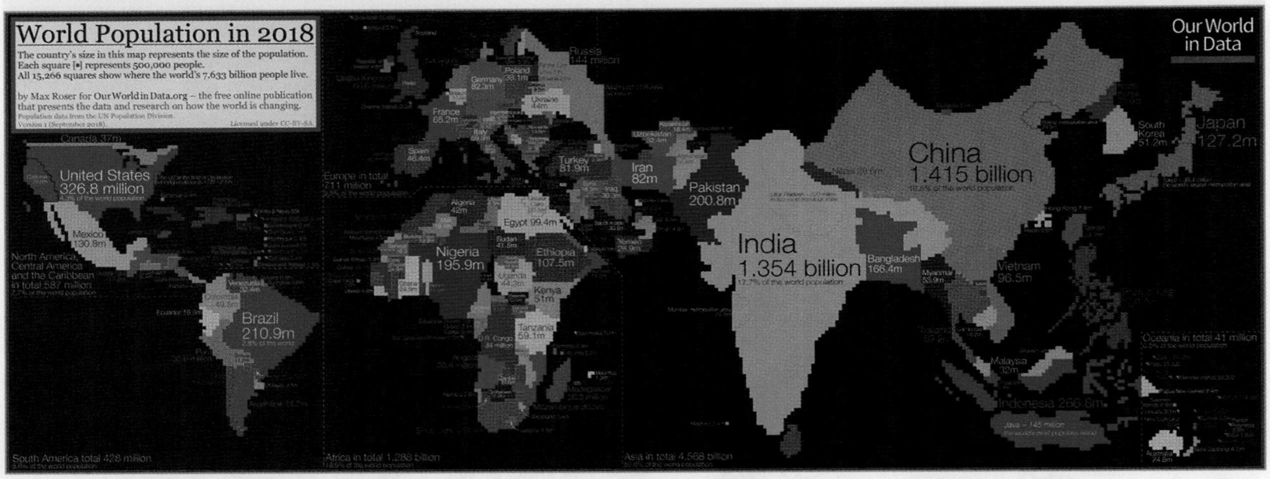

Source: © Our World in Data

Figure 3.3 shows the **population distribution** across Australia. The south-eastern and south-western parts of Australia are the most densely populated, and large areas of the continent are very sparsely populated, with fewer than 0.1 people per square kilometre. As you would expect, the highest population densities are located in and around the major cities — Sydney, Melbourne, Brisbane, Perth and Adelaide. In comparison, as you can see in figure 3.4, Europe is very densely populated, and only small sections of northern Europe have very low population densities.

In Australia, population densities are measured as the number of people per square kilometre. The formula for calculating **population density** is:

$$\frac{\text{total population}}{\text{total area}}$$

FIGURE 3.3 Australia's population density, 2017

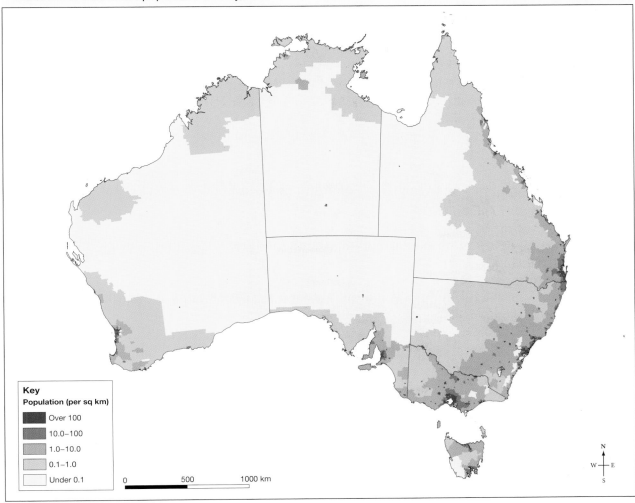

Source: Australian Bureau of Statistics

Australia's total area is 7 692 024 km² and its population in 2018 was 25 million. This means that Australia's overall population density in 2018 was 3.25 people per square kilometre.

$$\frac{25\,000\,000}{7\,692\,024 \text{ km}^2} = 3.25$$

In comparison, the area of Europe is 10 180 000 km² and its 2018 population was 742 648 010, so its population density was 72.95 people per square kilometre.

$$\frac{742\,648\,010}{10\,180\,000 \text{ km}^2} = 72.95$$

A knowledge of demography is important for an understanding of the potential needs of communities because the composition of populations involves a range of social and economic characteristics including age, sex, household structure, education levels, occupation, wealth and religion. Populations are also dynamic, so demography can be used by governments, organisations and individuals to identify potential social, economic and environmental issues for places and help find possible solutions to population-related problems. These needs and issues may occur at a variety of scales, ranging from local (e.g. a suburb or even your school), through regional (e.g. north Queensland) to national and global.

FIGURE 3.4 Europe's population density, 2017

Population density in Europe

■ Areas with 250 people or more (per sq km)
Source: Alasdair Rae / The Conversation

3.2.2 Demographic concepts

There are ten key demographic concepts you will need to know for this topic. These are:

Birth rate: the annual number of live births per 1000 people (also referred to as the crude birth rate). Crude birth rates for countries are calculated using the formula:

$$\text{births per 1000} = \frac{\text{births per year}}{\text{total population}} \times 1000$$

Death rate: the annual number of deaths per 1000 people (also known as crude death rate). The formula for calculating crude death rates is:

$$\text{deaths per 1000} = \frac{\text{deaths per year}}{\text{total population}} \times 1000$$

Fertility rate: the average number of children per woman of child-bearing age (usually ages 15 to 49) during her lifetime.

Life expectancy: the average number of years a newborn infant is expected to live, given the **mortality rates** at the time of their birth.

Age–sex structure: the composition of a population by age (e.g. 0–4 years, 5–9 years, etc) and sex (male, female).

Migration rate: immigration (incoming) number minus emigration (departing) number per 1000 people. **Net migration** is the number of immigrants minus the number of emigrants, as per the formula:

$$\text{net migration rate} = \text{immigration number} - \text{emigration number}$$

Total population growth: how much the population of a place has grown, taking into account births, deaths, immigration and emigration. It is determined using the formula:

$$\text{total population growth} = \text{population} + (\text{births} - \text{deaths}) + (\text{immigration} - \text{emigration})$$

Rate of natural increase: the birth rate minus the death rate, expressed as a percentage. The formula used to calculate this is:

$$(\text{crude birth rate} - \text{crude death rate}) \times \frac{100}{1000}$$

Percentage urban: the percentage of the population living in areas termed 'urban' (towns and cities) by that country or the UN.

GNI per capita ($US PPP): this stands for 'Gross National Income in Purchasing Power Parity' and is a measure of a country's relative wealth in current international dollars. It must be divided by the mid-year population.

Table 3.1 provides a statistical summary of these key demographic concepts for several countries, and figures 3.5, 3.6, 3.7 and 3.8 illustrate the global patterns of four of the indicators. The table also provides data for population size of each country.

TABLE 3.1 Demographic characteristics of four countries, 2018

Demographic feature	Australia	Niger	Japan	Italy
Population size (millions)	24.1	22.2	126.5	60.6
Birth rate (per 1000 people)	13	48	8	8
Death rate (per 1000 people)	7	10	11	11
Rate of natural increase (%)	0.6	3.8	−0.3	−0.3
Total fertility rate (average per female aged 15–49 years)	1.7	7.2	1.4	1.3
Life expectancy [male, female] (years)	82 [80, 85]	60 [59, 61]	84 [81, 87]	83 [81, 85]
Population under 15 years of age (%)	19	50	13	13
Population over 65 years of age (%)	16	3	28	23
Net migration rate (per 1000 people)	10	0	1	3
Percentage urban	86	16	92	70
GNI per capita ($US PPP) (in 2017)	45 780	990	38 260	40 030

Source: Toshiko Kaneda, Charlotte Greenbaum, and Kaitlyn Patierno, 2018 World Population Data Sheet Washington, DC: Population Reference Bureau, 2018. Reproduced by permission. All rights reserved.

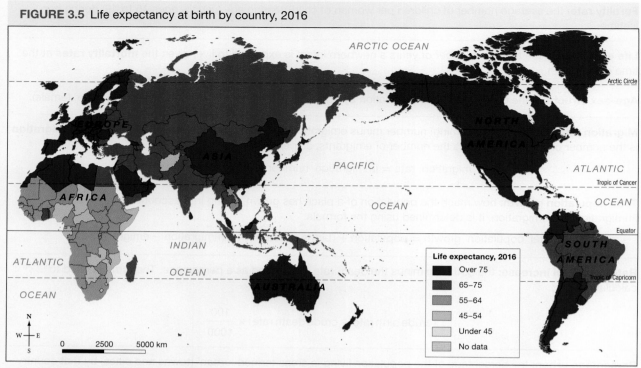

FIGURE 3.5 Life expectancy at birth by country, 2016

Source: Central Intelligence Agency

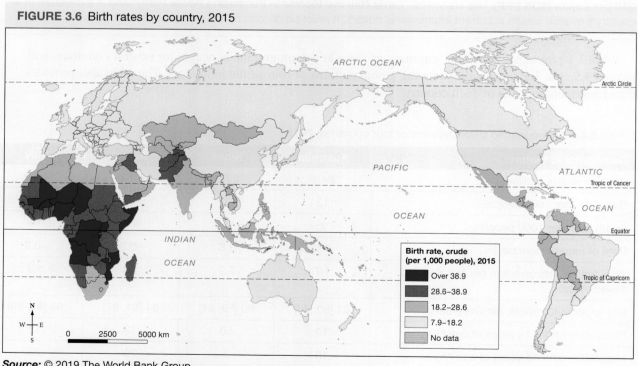

FIGURE 3.6 Birth rates by country, 2015

Source: © 2019 The World Bank Group

Table 3.1 and figures 3.5, 3.6, and 3.7 and 3.8 show that there are significant differences between countries and regions around the world in terms of **demographics**. For example, the current average birth and death rates for **more developed countries** are 10 per 1000 for both rates, while those for the **least developed countries** are 33 for birth rate and a death rate of 8.

FIGURE 3.7 Fertility rate by country, 2015

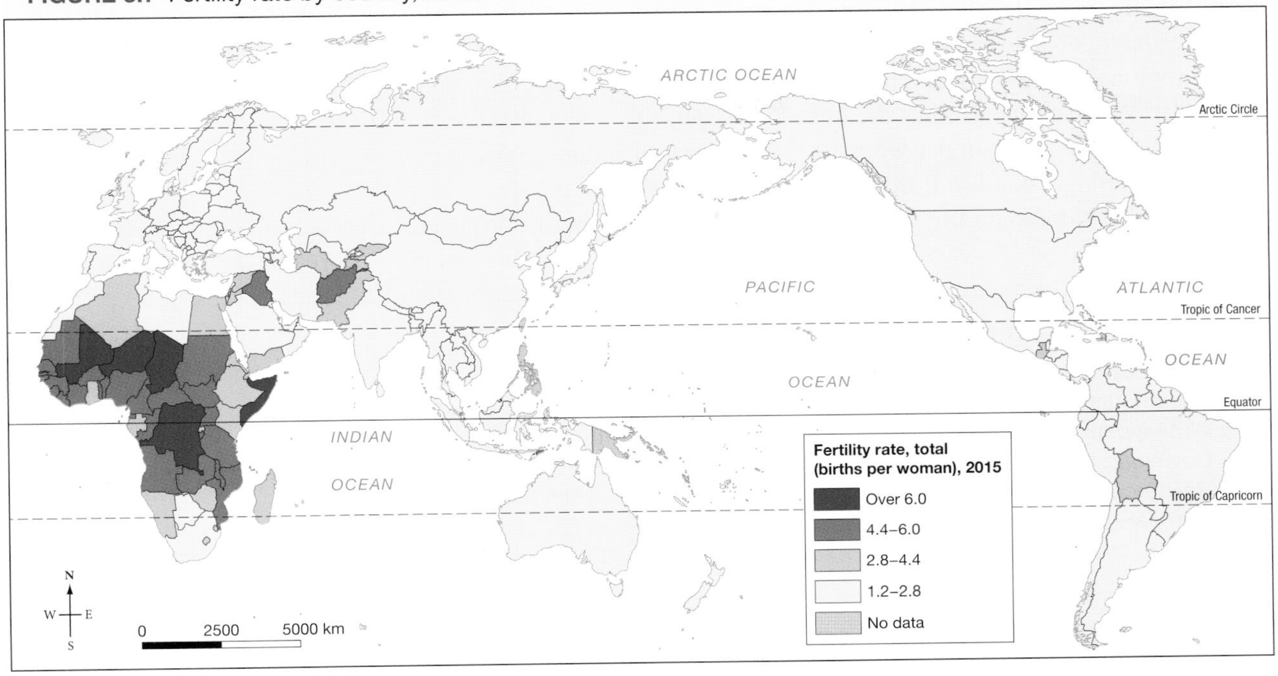

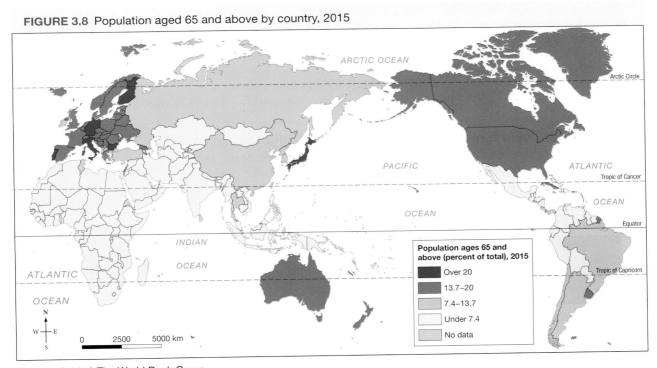

Source: © 2019 The World Bank Group

FIGURE 3.8 Population aged 65 and above by country, 2015

Source: © 2019 The World Bank Group

For fertility rates, the figures are 1.6 children on average for developed countries and 4.2 children on average for the least developed countries. The life expectancy figures are 72 years for most developed countries compared to 65 years for least developed countries. More developed countries have a much smaller percentage of their population below 15 years of age (16 per cent compared to 40 per cent in the least developed countries) and a greater percentage above 65 years of age (18 per cent compared to four per cent). Finally, more developed countries have a positive net migration rate, while the least developed countries have, on average, more emigrants than immigrants.

The regions with the highest birth rates around the world are West and Middle Africa (38 and 42 births per 1000). West Africa also has the lowest average life expectancy (57 years) and Middle Africa the highest fertility rates (5.9 children per woman). The lowest birth rates, highest life expectancies and lowest fertility rates are generally found in Europe. In southern Europe, birth rates are around eight per 1000, fertility rates are 1.4 children and life expectancy is 82 years. Southern Europe also has the lowest percentage of its population below 15 years (14 per cent) and the highest population above 65 years (20 per cent).

Activity 3.2a: Comparing demographic statistics
Explain and analyse the data
1. Use the data in table 3.1 to describe the key features of Australia's demography in 2018.
2. Write a paragraph to compare Australia's demography with that of Japan, Italy and Niger.
3. Download the latest World Population Data Sheet from the **Population Reference Bureau** website in the Resources tab. Draw up a table similar to table 3.1 for a selection of countries.
4. Compare the demographic features of your selected countries with those of Australia.

Activity 3.2b: Describing and comparing demographic patterns
Explain and compare patterns of data
1. Describe the patterns for each of the demographic indicators illustrated in figures 3.5, 3.6, 3.7 and 3.8. Use the various regions used by the Population Reference Bureau on its World Population Data Sheet as the basis of your description (Northern Africa, Western Africa, Eastern Africa, etc.)
2. Compare the regional patterns of life expectancy, birth rates, fertility rate and population aged 65 and above. Suggest reasons for similarities and differences in these demographic indicators.
3. Identify any connections you think there might be between the three indicators. For example, is life expectancy connected to birth rates?
4. Explain any connections you identified in question 3, if you think there are any.

Activity 3.2c: Using formulas to calculate demographic measures
1. Use the formulas for calculating population density, crude birth and death rates, and the rate of natural increase to complete table 3.2.
2. Use your calculations to compare these demographic features of Australia's states. Look for similarities and differences.

TABLE 3.2 Demographic features of Australia's states

State	Population 2017	Area (km^2)	Population density (persons per km^2)	Births 2017	Crude birth rate	Deaths 2017	Crude death rate	Rate of natural increase
New South Wales	7 867 936	800 642		96 591		52 778		
Victoria	6 321 606	227 416		82 094		39 791		
Queensland	4 927 629	1 730 648		61 158		31 555		
Western Australia	2 574 193	2 529 875		34 498		14 494		
South Australia	1 723 923	983 482		19 072		14 052		
Tasmania	522 410	68 401		5610		4780		

Source: Australian Bureau of Statistics

3.2.3 Age–sex structure and population pyramids

Table 3.1 provided data on the age structure of the selected countries based on the percentage of population below 15 years and above 65 years of age. **Population pyramids**, (also known as **age–sex pyramids** or population profiles), show more detailed data on the age of a country's population, as well as the male–female breakdown of the population. The 2019 population pyramids for Australia, Niger and Italy are shown in figures 3.9, 3.10 and 3.11.

Population pyramids are divided horizontally into two parts: the left side for the male population and the right side for the female population. The scale for these two sections of the graph can be population numbers or percentage of population (see figure 3.12). Whichever scale is used, the divide will remain the same. Within the two parts of the pyramid, the population is then divided vertically into 5-year age cohorts (0–4 years, 5–9 years, etc).

The demographic characteristics of countries as illustrated by population pyramids can be divided into three broad categories: expansive, stationary and contractive. Expansive populations such as Niger's have larger numbers in the younger age groups and progressively smaller numbers in older age groups. Expansive populations have relatively high birth and fertility rates, and lower than average life expectancies. Because of the younger age of the overall population, death rates are usually low. Niger has a typical expansive population pyramid shape, with a broad base and narrow top.

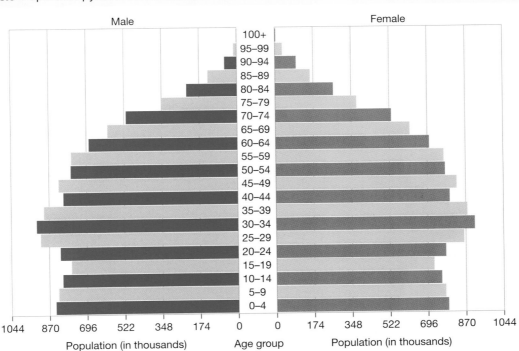

FIGURE 3.9 Population pyramid for Australia, 2019

Source: United Nations, Department of Economic and Social Affairs, Population Division 2017. World Population Prospects: The 2017 Revision, custom data acquired via website.

FIGURE 3.10 Population pyramid for Niger, 2019

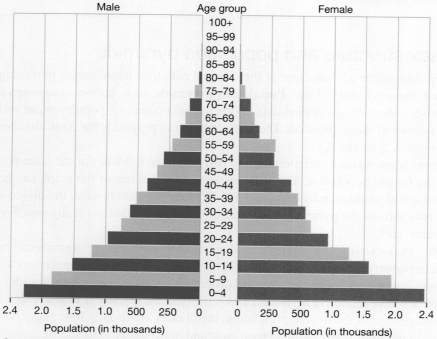

Source: United Nations, Department of Economic and Social Affairs, Population Division 2017. World Population Prospects: The 2017 Revision, custom data acquired via website.

FIGURE 3.11 Population pyramid for Italy, 2019

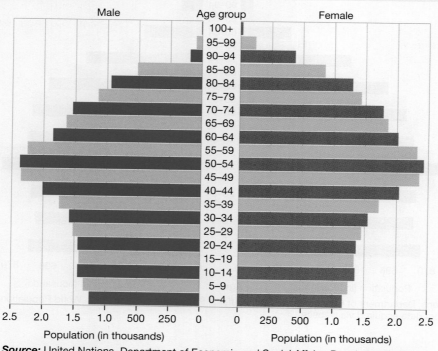

Source: United Nations, Department of Economic and Social Affairs, Population Division 2017. World Population Prospects: The 2017 Revision, custom data acquired via website.

FIGURE 3.12 World population pyramid, 2019

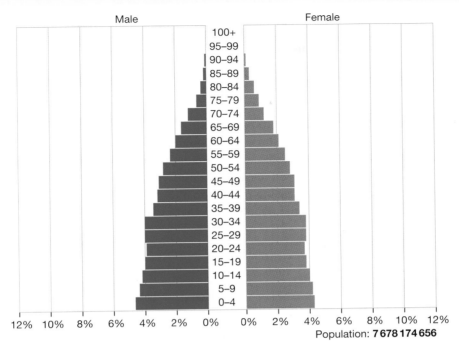

Source: PopulationPyramid.net

In contrast, contractive populations such as those of Italy have relatively smaller numbers in the younger age groups and larger numbers in the older age groups. Contractive populations have low birth and fertility rates, and high life expectancies. Because of their ageing population, death rates are commonly higher than those in expansive populations, although Italy's death rates are still low when compared with other countries. Countries with a higher death than birth rate may have a declining population, although a positive net migration may compensate for a negative rate of natural increase. The demographic processes involved in population change, including the role of migration, are covered in subtopic 3.3.

Activity 3.2d: Age–sex structure and population pyramids

Refer to table 3.1 and figures 3.9, 3.10, 3.11 and 3.12 to answer the following questions.

Explain and analyse the data

1. Explain how the shape of each country's pyramid reflects the birth and death rates and average life expectancy for that country.
2. Given the shape of the world population pyramid, what would you estimate to be the birth and death rates and average life expectancy globally? Give reasons for you answer.
3. Which age cohorts had the largest numbers of males and females in each of the countries? What was the population in each case?
4. The populations of Australia, Niger and Italy in 2019 were approximately 25.3, 23.3 and 58.8 million respectively. What percentage of the total population of each country was found in the most populous age cohorts?
5. Identify the age cohorts that are used to calculate the fertility rate for each country. Explain how the shape of each pyramid reflects the fertility rate of each country.
6. Which of the individual country pyramids most closely resembles the world population pyramid? Why might that be the case?
7. Use the **US Census Bureau** weblink in the Resources tab to locate the population pyramids of several other countries. Write a description of the demographic features of each country illustrated by its pyramid.

> **Resources**
>
> 🔗 **Weblink** US Census Bureau

3.3 Changes in population

3.3.1 Introduction

In 2018, the world's population was estimated to be 7.6 billion, with China (1 394 million), India (1 371 million), and the US (328 million) the three most populous countries. In the 1960s, the total world population was around 3 billion. This means that world population more than doubled in 50 years, and will continue to grow into the future, possibly even reaching 10 billion by the mid-2050s. Table 3.3 shows that the world's population grew at an ever-increasing rate in the mid- to late-20th century. What will the world look like when you are 50? (This exponential population growth will be examined in more detail in topic 4.)

TABLE 3.3 Global population growth

Year	1800	1930	1960	1974	1987	1999	2011	2024
Population	1 billion	2 billion	3 billion	4 billion	5 billion	6 billion	7 billion	8 billion

Source: © 2019 Population Reference Bureau, United Nations, Our World in Data

Australia's population has also more than doubled since the 1960s, although its relative contribution to global population growth is relatively insignificant. In the 1960s, Australia's population was approximately 11 million; by late 2018, the population had reached 25 million, so about 14 million people had been added to Australia's population in about 50 years. In comparison, India added more than 900 million people to its population over the same period (see table 3.4 on page 133). Use the weblinks in the Resources tab for introductory videos and interactivities about population concepts.

> **Resources**
>
> 🔗 **Weblinks** Birth rates
> Population concepts
> Demographic indicators (Adobe Flash needed)
> Population growth

3.3.2 Rates of natural increase and decrease, and population change

World population and the populations of most countries have continued to grow over time because the number of babies born each year has exceeded the number of people who die. That is, birth rates have been greater than death rates. This means that the rate of natural increase has been positive. Not all countries have a positive rate of natural increase, including Italy and Japan. Nevertheless, countries where there is a natural decrease in population are far outweighed by countries whose populations are increasing.

Longevity has also played a role in the growing populations of many countries, although over time, given declining fertility rates, this will become less of a factor in population growth. Average life expectancy in the 1960s was around 53 and is now about 72 (70 for males and 74 for females). Large numbers of people born in the 1960s are still alive today and have had families, so longevity has been part of the growth in global population.

Changes over time in birth rates and death rates have an impact on the rate at which a country's population changes. If birth rates are much higher than death rates for a lengthy period, the population will grow rapidly; if the death rate exceeds the birth rate, the population will, in most cases, decline. If the rates are close, the population will change slowly.

Changes in birth rates and death rates over time are commonly shown as line graphs, with one line for birth rates and a second line for death rates. The gap between the lines indicates the extent to which the population is growing or declining (the rate of natural increase or decrease). Figure 3.13 is an example of a line graph for the world's birth and death rates between 1960 and 2016.

FIGURE 3.13 Global birth and death rates, 1960–2016

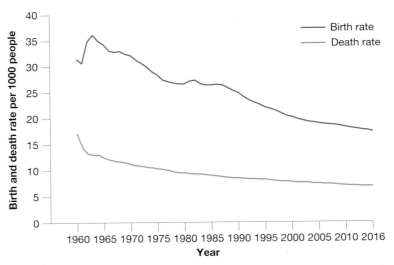

Source: © 2019 The World Bank Group

Activity 3.3a: Explaining rates of natural increase and decrease and population change

Refer to figure 3.14 to answer the following questions.

FIGURE 3.14 Rate of natural increase by country (%), (a) 1960, (b) 1990 and (c) 2015

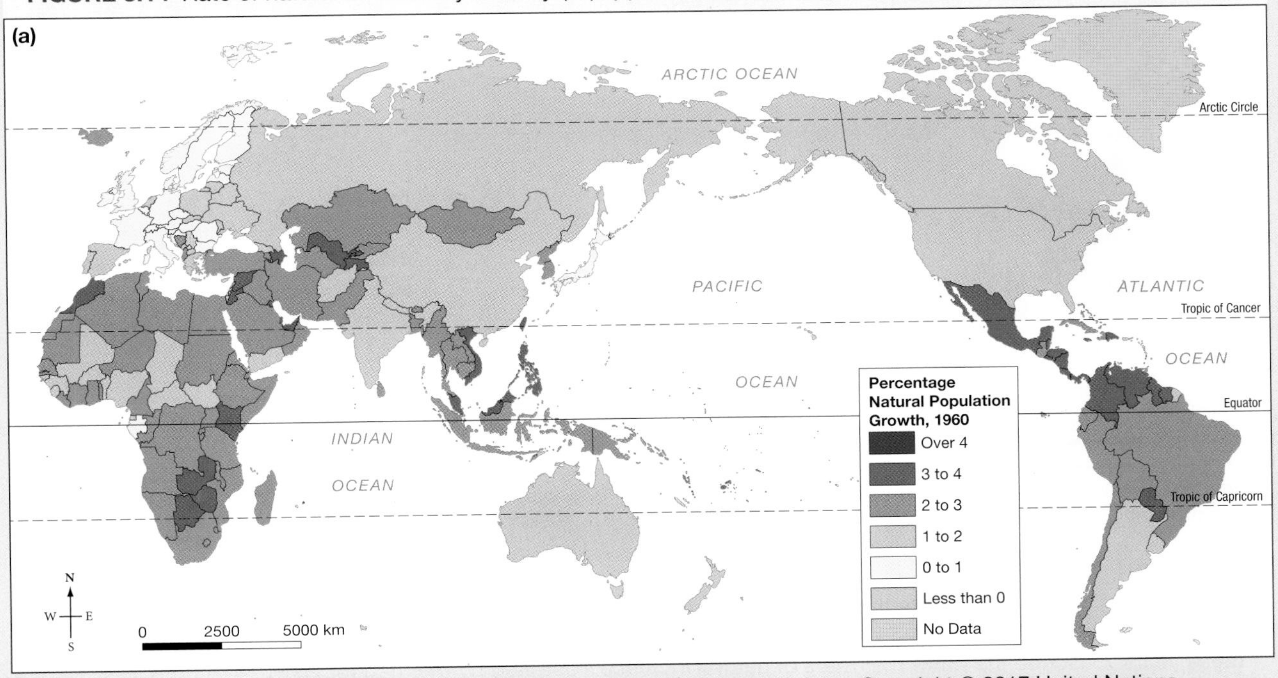

Source: From World Population Prospects 2017 by United Nations / Population Division. Copyright © 2017 United Nations. Used with permission of the United Nations.

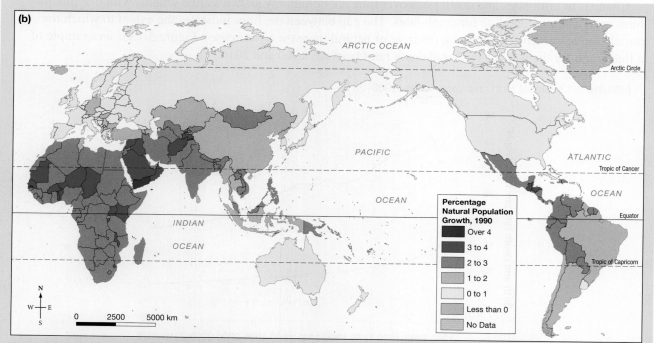

Source: From World Population Prospects 2017 by United Nations / Population Division. Copyright © 2017 United Nations. Used with permission of the United Nations.

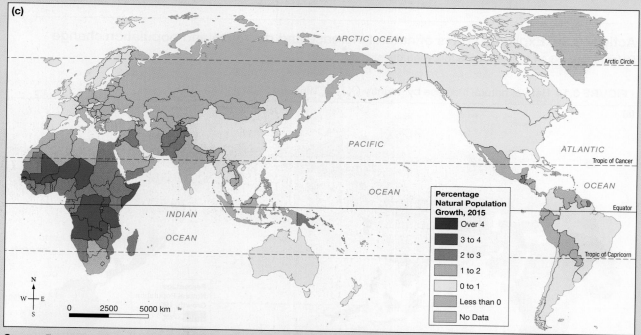

Source: From World Population Prospects 2017 by United Nations / Population Division. Copyright © 2017 United Nations. Used with permission of the United Nations.

Explain and analyse data

1. The maps in figure 3.14 illustrate the spatial distribution pattern of natural population growth in 1960, 1990 and 2015. Different countries and regions of the world have experienced different rates of growth in each of these years. Describe the pattern of rates of natural increase illustrated by the map for 2015. How does this map provide evidence that the world's population is continuing to grow?
2. Which regions of the world are experiencing the highest and the lowest rates of natural increase? Suggest reasons for why this might be the case.

3. Identify some of the countries at the various rates of growth used for the key to the 2015 map. Create a table to show the rates of natural increase and the country examples. You should include Australia as one of the countries on the table.
4. Describe the spatial distribution pattern of changes that have occurred in rates of natural increase from 1960 to 1990 and 2015. Refer to regions and specific countries in your answer. Identify the countries where rates have changed the most.
5. You can examine the changing patterns of rates of natural increase around the world using the interactive maps provided by the websites **Our World in Data** and **Knoema** in the Resources tab. You may want to use these maps to describe the changes over time in more detail for particular regions or countries.

Resources

Weblinks Our World in Data
Knoema

Activity 3.3b: Identifying population trends over time

Refer to figure 3.13 on page 131 and table 3.4 to answer the following questions.

TABLE 3.4 Birth and death rates and population change in Australia and India

Year	Australia				India			
	Birth rate (per 1000)	Death rate (per 1000)	Rate of natural increase (%)	Total population (millions)	Birth rate (per 1000)	Death rate (per 1000)	Rate of natural increase (%)	Total population (millions)
1891	34.5	16.6	1.79	3.2	48.9	41.3	0.76	235.9
1901	27.2	12.1		3.6	45.8	44.1		238.4
1911	27.2	10.8		4.6	49.2	42.6		252.1
1921	25.0	9.8		5.5	48.1	47.2		251.3
1931	18.2	8.5		6.5	46.4	36.2		279.0
1941	18.9	10.0		7.1	45.9	37.2		318.7
1951	23.1	9.6		8.5	39.9	27.4		361.1
1961	21.3	8.7		10.6	41.7	22.8		439.2
1971	18.8	8.3		12.6	41.2	19.0		548.2
1981	15.8	7.6		15.0	37.2	15.0		683.3
1991	14.9	6.9		17.4	29.5	9.8		846.3
2001	12.7	6.6		19.4	25.4	8.4		1033.4
2011	13.3	6.5		21.5	21.8	7.1		1210.2
2018	13.0	7.0		24.7	20.0	6.0		1371.3

Source: Australian Bureau of Statistics, © Office of the Registrar General & Census Commissioner, India

Explain and analyse the data

1. Describe what has happened over time to the world's birth and death rates.
2. What does the gap between the birth and death rate lines tell us about global population change over time?
3. The rate of natural increase is the birth rate minus the death rate, expressed as a percentage. Calculate the rates of natural increase for Australia and India and add these figures to table 3.4.

4. In which years were the birth and death rates and the rate of natural increase highest and lowest in each country?
5. Describe how the rates of natural increase have changed over time in Australia and India. How might these changes relate to population changes in both countries? In the case of Australia, what other factors need to be taken into account to explain the growth in population?
6. Draw a multiple line graph for birth and death rates in Australia and India. Add a scale for population on the right hand side of the graphs, and draw a line to show the change in population over time for each country.
7. Describe the pattern of population change illustrated by your completed graph. In particular, look at the gap between the birth and death rates and at the line showing population change.
8. Compare population change over time in Australia and India. Is there a constant downward trend in both birth and death rates? Have there been upward spikes in either of the rates? If there have been spikes, what might account for them?

Resources

Video eLesson Skillbuilder: Constructing multiple line and cumulative line graphs (eles-1740)
Interactivity Skillbuilder: Constructing multiple line and cumulative line graphs (int-3358)

3.3.3 Changes in age–sex structure

Population pyramids are used to show the age–sex features of countries' populations, but the structure and shape of population pyramids can also be used to identify demographic features such as birth and death rates, fertility rates and life expectancy. Changes over time in countries' pyramids illustrate changes in these demographic characteristics. Figure 3.15 provides a series of population pyramids for Australia.

FIGURE 3.15 Population pyramids for Australia, (a) 1961, (b) 1991 and (c) 2018

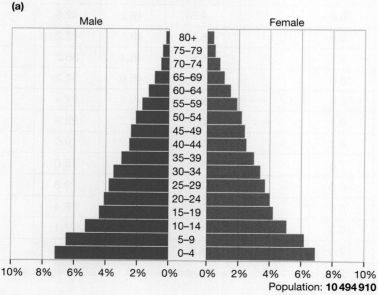

Source: PopulationPyramid.net

(b)

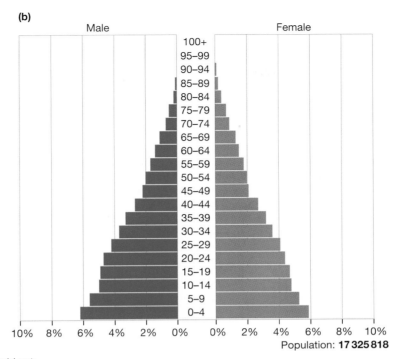

Population: **17 325 818**

Source: PopulationPyramid.net

(c)

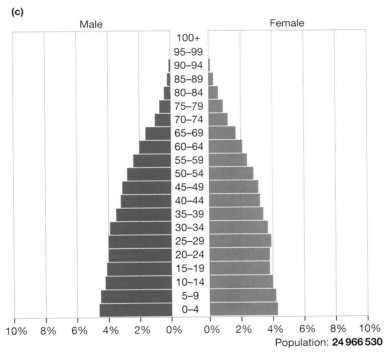

Population: **24 966 530**

Source: PopulationPyramid.net

The changes illustrated by the pyramids often have implications for planning by governments, non-government organisations, businesses and individuals. For example, a relatively young population would indicate a need for more schools and housing in the future, while a relatively older population would need increased health and aged-care facilities.

Activity 3.3c: Explaining changes in age–sex structure of Australia's population

Refer to figure 3.15 to answer the following questions.

Explain and analyse population data

1. Identify the age cohorts (male and female) with the highest percentage of the population in the years 1961, 1991 and 2018. How have these age cohorts changed over time?
2. Describe the changes that have occurred in Australia's population pyramids over time.
3. What do these changes tell us about how Australia's demographic characteristics (birth and death rates, fertility rates, rate of natural increase, life expectancy) have changed over time? Refer to the data in table 3.4 to support your answer to this question.
4. Use the **ABS animated population pyramids** website in the Resources tab to look at Australia's predicted 2040 population pyramid. Compare the demographic features illustrated by this pyramid with that of 2018. How is Australia's population expected to change over time? What might be some of the implications of these changes?
5. What predicted changes will occur in the 15–19 cohort by 2040? What implications will this have on you?

Resources

Weblink ABS animated population pyramids

3.4 Factors affecting population change

3.4.1 Introduction

Populations change both naturally and through migration. These changes are influenced by a combination of economic, social, cultural, political and epidemiological (public health) factors that include:

- advances in health care and life expectancy
- birth rates and the changing role of women in society
- the impact of disease on death rates
- migration policies over time
- **amenity**.

These factors are often linked. For example, as women's roles in Australian society changed from the 1960s, birth and fertility rates declined (see table 3.4). Improved health care has resulted in both improved life expectancy and declining death rates. This subtopic will examine factors that have affected Australia's changing population.

FIGURE 3.16 Factors influencing population change

- Advances in healthcare and life expectancy
- Changing birth rates and role of women in society
- Migration policies over time
- Amenity
- Impact of disease on death rates
- Population change

3.4.2 Factors affecting natural increase

The impact of disease on death rates

Death rates have declined over time in Australia, from over 16 per 1000 people in the 1890s to around 7 per 1000 today. In fact, between 1907 and 2016, the age-standardised death rate in Australia fell by 74 per cent. Figure 3.17 illustrates one important factor in the fall in Australia's death rates. In the early part of the 20th century, the second most common cause of deaths in Australia was infectious diseases, such as tuberculosis, typhoid, diphtheria, measles, scarlet fever and pneumonia. These were responsible for a quarter of all deaths in Australia, many of them children and young adults. However, by the 1950s, these diseases had become much less common and were responsible for relatively few deaths.

> Age-standardised death rates are used to compare death rates over time. The Australian Bureau of Statistics defines age-standardised rates as 'hypothetical rates that would have been observed if the populations being studied had the same age distribution as the standard population, while all other factors remained unchanged.'

There is a link between the control of infectious diseases in Australia and the decline in death rates over time. If you compare the death rate line of your graph in activity 3.3b with that for infectious disease in figure 3.17, you will see that they closely correspond.

Factors such as control of infectious disease, better hygiene and better nutrition were responsible for the decline in deaths from infectious diseases. By the 1950s, immunisation against infectious diseases was almost universal in Australia, as was the use of recently discovered antibiotics such as penicillin.

As deaths from infectious diseases decreased in the 1940s, deaths from circulatory diseases (heart attacks, strokes, etc) increased sharply and became the leading cause of death in Australia. By the 1970s, improvements in the detection and treatment of cardiovascular disease, as well as improvements in its

FIGURE 3.17 Age-standardised death rates by broad cause of death, 1907–2016

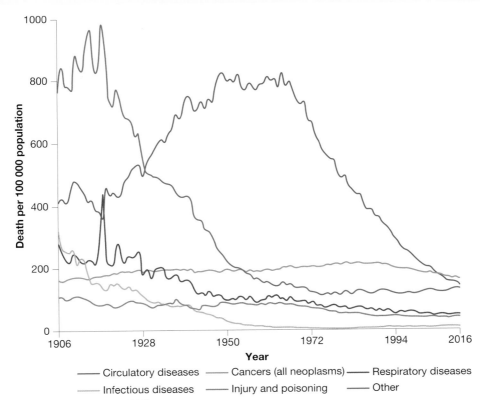

Source: Australian Institute of Health and Welfare.

prevention, meant that the number of deaths from circulatory diseases fell sharply. Falling rates of smoking also contributed to the decline. Figure 3.18 illustrates how dramatic the decline in deaths from heart and cerebrovascular disease was in Australia between 1968 and 2017.

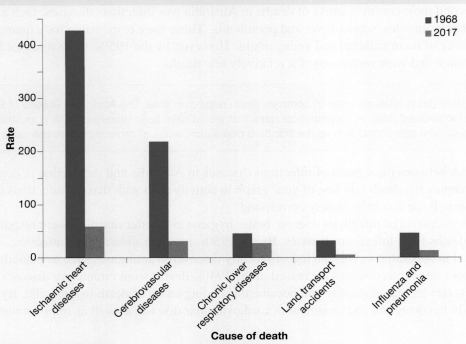

FIGURE 3.18 Standardised death rates for top five leading causes of death in 1968 and 2017

Source: Australian Bureau of Statistics

While death rates have gradually declined over time in Australia, there have occasionally been upward spikes. Table 3.4 illustrated one spike in the 1940s, an indication of the impact of World War II on Australia. Another spike occurred in 1918–19 as a consequence of a virulent strain of influenza that affected most of the world. It is estimated that this 'Spanish flu' pandemic was responsible for over 15 000 deaths in Australia over a 6-month period (see the spike in respiratory diseases on figure 3.17). Unlike most outbreaks of influenza, more than 50 per cent of deaths from Spanish flu occurred in people aged between 20 and 39 years, rather than infants or older people, and was more common in males than females. Despite occasional upward spikes, better management and control of disease over time have led to declining death rates in Australia.

Advances in health care and life expectancy

Improved disease control not only led to a reduction in Australia's death rates over time, it also contributed to dramatic improvements in life expectancy (see figure 3.19). Girls and boys born in 2016 can expect to live 34 and 33 years longer respectively than the babies who had been born in the late 19th century. Advances in health care have also contributed to increased life expectancy and the decline in death rates in Australia.

Advances in health care involve a range of factors. Advances in medical technology and treatment techniques, improvements in health care infrastructure such as hospitals and aged-care facilities, and high quality medical training and research have all helped. Even public health campaigns designed to reduce rates of smoking and warn about the risks of skin cancer form part of Australia's advances in health care. Figure 3.20 shows the range of demographic, social, economic and political factors involved in Australia's current health care system.

FIGURE 3.19 Life expectancy at birth, Australia, 1891–2015

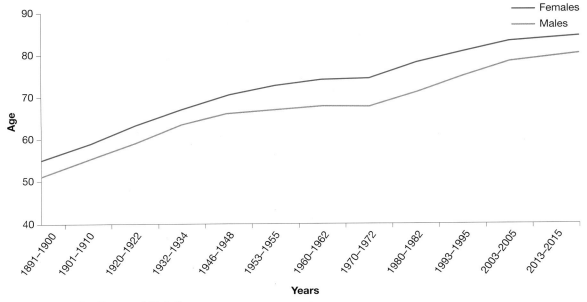

Source: Australian Bureau of Statistics

FIGURE 3.20 Australia's health landscape

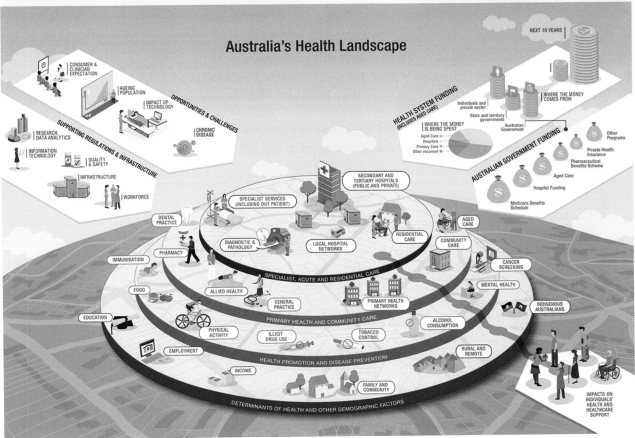

Source: © Gravity Consulting and Gravity iLabs

FIGURE 3.21 Australian health care, past and present: (a) Sydney Hospital surgery, 1914, (b) hospital surgery, 21st century, (c) hospital ward, late 19th century, (d) hospital ward, 21st century

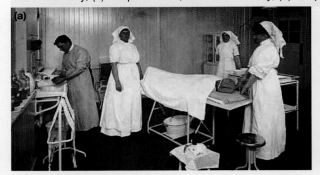
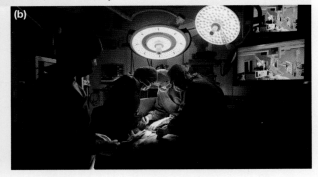

Source: NSW State Archives

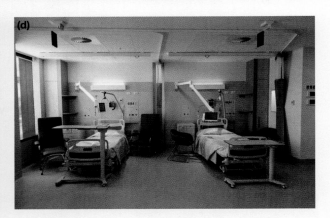

Source: Mitchell Library, State Library of New South Wales

Birth rates and the changing role of women in society

Birth rates in Australia have fallen over time, from a high of 34.5 in 1890 to a low of 12.7 in 2001. Overall, the trend has been downwards, although there have been more fluctuations in birth rates than in death rates. For example, after a low point in 2001, birth rates rose to 13.3 in 2011, and still remained higher in 2018 than in 2001. The most pronounced upward movement in birth rates occurred after World War II. This is known as the **baby boom** period. Birth rates were higher in the 1950s and 60s than they had been since the 1920s.

There are a number of social, economic and cultural factors that help explain changes in birth rates in Australia over time. They include:
- the changing role of women in society, including the increased participation of women in the workforce
- the availability of family planning techniques, including contraceptives
- an increase in the age at which women marry
- the use of government incentives such as baby bonuses
- increases in the compulsory schooling age
- increases in the costs associated with children, such as the cost of education
- reduced infant and childhood mortality rates
- increased **urbanisation** and the reduction in the number of farming families
- social disruption, especially that caused by war.

Some demographers consider that the most important factor for the decline over time in birth rates in Australia and other countries with very high levels of development, is the changing role of women in society. As educational opportunities for girls increased and more women entered and remained at work, women chose to marry later and to delay having children. On average, child-bearing is now occurring an average of ten years later compared to the 1950s, so many women now have their first babies in their 30s rather than their 20s. In addition, many women have chosen to have only one or two children, while others, perhaps up to 25 per cent, have chosen to have none at all. The impact of these decisions on Australia's fertility rates is shown in figure 3.22. The graph suggests that the most dramatic changes in the role of women in Australian society occurred in the 1960s and 70s, when fertility rates declined rapidly.

Perhaps the most important change in the role of women in Australian society has been in their increased participation in the workforce. Before the 1960s, most women remained at home and cared for their children. However, from the 1960s, as society changed culturally, socially and economically, increasing numbers of women entered the workforce. Female workforce participation rates have increased by 50 per cent over the past 40 years.

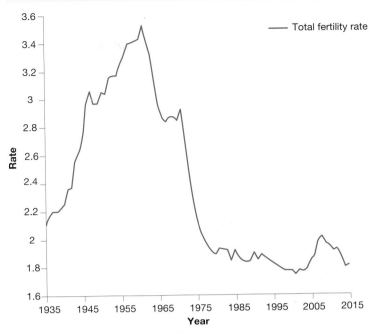

FIGURE 3.22 Total fertility rate for Australia, 1935–2015

Source: Australian Bureau of Statistics

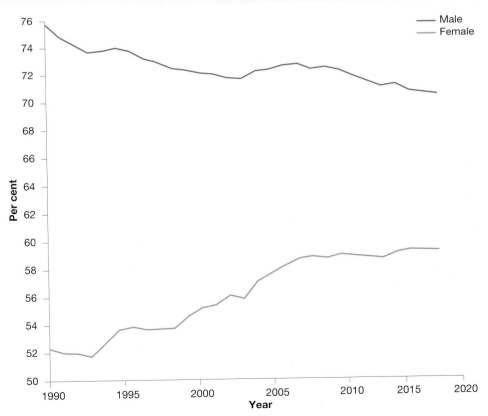

FIGURE 3.23 Male and female workforce participation rates, 1990–2018

Source: © 2019 The World Bank Group

TABLE 3.5 Birth rates, fertility rates and female workforce participation rates, 1960–2018

Year	Birth rate (per 1000 people)	Fertility rate	Workforce participation rate (women aged 25–64)
1960	22.4	3.45	29.1 (est)
1965	19.6	2.98	34.3
1970	20.6	2.86	39.5
1975	16.8	2.15	44.6
1980	15.3	1.89	46.48
1985	15.7	1.92	49.42
1990	15.4	1.90	59.5
1995	14.2	1.82	61.7 (est)
2000	13.1	1.77	63.9
2005	13.1	1.87	68.08
2010	13.8	1.95	70.29
2015	13.1	1.81	71.68
2017	12.8	1.74	72.93
2018	12.7	1.70	N/A

Source: Australian Bureau of Statistics, © Statista 2019

3.4.3 Factors affecting migration rates over time

Net overseas migration to Australia has fluctuated a great deal over time, both in terms of the actual numbers of immigrants and the relative role immigration has played in Australia's overall population growth. The migration of people within Australia has also changed a great deal both over time and spatially. Table 3.6 provides data on net interstate migration between 1975 and 2017. The table shows that New South Wales has consistently lost people to other parts of Australia, although overseas immigration and natural increase has more than compensated for New South Wales's negative net interstate migration rate. Queensland is the only Australian state or territory that has consistently had more people migrating from other parts of Australia than have departed.

TABLE 3.6 Net interstate migration, 1975–2017

	NSW	VIC	QLD	SA	WA	TAS	NT	ACT
1975	−20 347	−15 356	11 326	3239	8144	−509	8636	4866
1980	−6401	−13 604	23 895	−6066	1601	−1003	1976	−398
1985	−12 232	−10 290	14 217	−1899	5730	261	1026	3187
1990	−27 923	−9053	32 549	515	598	2169	−637	1782
1995	−14 441	−18 205	38 472	−7845	4513	−2731	1416	−1179
2000	−14 708	4920	20 367	−3699	−2501	−2533	−1621	−218
2005	−26 484	−3571	29 141	−3366	2818	306	726	470
2010	−10 849	3131	5384	−3038	4457	714	−1599	1740
2015	−11 539	17 639	11 986	−7212	−10 010	760	383	−2029
2017	−15 161	18 193	17 795	−6778	−13 934	1522	−2867	1230

Source: Australian Bureau of Statistics

Migration rates, whether international, **inter-state** or **intra-state** (within states and territories), are influenced by a range of economic, social, cultural, environmental and political factors. Unlike rates of natural increase, political factors often play an important role in migration rates, especially Australia's international immigration rates.

The role of amenity in rates of migration

The term amenity refers to the desirable or useful features or facilities of places. According to the Australian Government's 2016 *State of the Environment* report, amenity means access to shops and other services required for daily living, including access to employment, health care, educational services, transport, cultural and leisure services, and green spaces. Amenity is often a key component in the decision made by people to migrate from one place to another. Places with higher amenity are often attractive to potential migrants, and people may find the low levels of amenity in some places a factor in their decision to leave.

The factors that influence a person's decision to move from one place to another are referred to as **push and pull factors**. The positive features of places that are attractive to migrants are pull factors — they draw people towards places. In contrast, push factors are those negative features of places that are responsible for people leaving. Pull factors are features associated with a migrant's destination, while push factors are features of the migrant's place of origin.

Pull factors include things such as better employment opportunities, better access to educational and health services, a pleasant natural environment and so on. Push factors may involve lack of employment and services, and environmental disasters such as drought. In addition to these amenity factors, a number of Australia's international migrants are pushed away from their place of origin by conflict and persecution. These people form part of Australia's humanitarian and refugee migrant intake.

Migration policies over time

Unlike inter- and intra-state migration, which are largely determined by push and pull factors, the level and composition of Australia's overseas immigration are determined more by changing migration policies over time than by amenity. Overseas migrants are influenced by the attraction of Australia as a destination as well as by possible negative features of their places of origin, but the number of international migrants arriving in Australia is directly controlled by the federal government. This in large part explains the fluctuations that occur in the number of migrants arriving each year.

Currently, there are two programs for immigrants to Australia: a migration program and a humanitarian program. Both have two components: special skills and family for the migration program, and refugees and special humanitarian for the humanitarian program. In 2017–18, Australia received 128 550 special skill migrants, 57 400 family migrants, and 16 250 refugee and humanitarian migrants. This shows that the current migration policy has a focus on the skills migrants can offer to the Australian economy.

Decisions made by Australian governments about levels of migration and the types of migrants who can settle in Australia have been determined by a range of factors. For example, in the period after Federation in 1901, Australia's Immigration Restriction Act established what became known as the White Australia policy, which aimed to prevent non-Europeans from entering the country. Consequently, the vast majority of immigrants came from the United Kingdom and Ireland during this time. This focus on European immigrants only began to change in the 1960s, before finally ending in 1972.

One of the most significant changes in the numbers of migrants allowed into Australia occurred as a consequence of World War II. During the war, Australia had been threatened by the prospect of a Japanese invasion. At the end of the war, there was a wide-spread belief that Australia needed to 'populate or perish'. Not only was it thought that more people were needed for Australia's defence, immigrants were also needed to ensure high rates of economic growth and prosperity. Large numbers of non-British migrants and refugees began to arrive in Australia during the post-war period. Most came from war-torn Europe as displaced persons unable or unwilling to return to their home countries after the war, and many contributed to the workforce for large-scale infrastructure developments such as the Snowy Mountains Hydro-Electric Scheme. In total, more than 2 million immigrants settled in Australia between 1945 and 1966, nearly

171 000 displaced people settled in Australia in the 1940s and 50s, and the total population grew from 7.5 million to 11 million.

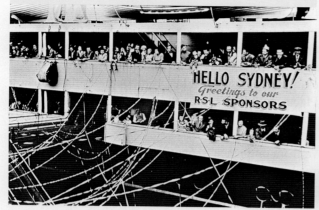

FIGURE 3.24 Migrants arrive in Australia post-World War II

FIGURE 3.25 Migrants working on the Snowy Mountains Scheme

Source: 1947, National Archives of Australia

A second refugee re-settlement program occurred in the period between 1976 and 1982, following the end of the war in Vietnam and the arrival of Indochinese 'boat people' in northern Australia. In response, the Australian government of the time developed the country's first planned Humanitarian Program. Indochinese refugees from camps in south-east Asia and Hong Kong were brought to Australia under this program. More recently, Australia has accepted refugees from war-ravaged Middle Eastern countries through its humanitarian program.

Australia's migration policies have caused major changes in both the size and composition of Australia's population. As well as now having a much larger population because of immigration, Australia has also become a multicultural society. Not all people agree with Australia's current migration policies, especially with respect to the numbers of migrants and the treatment of **asylum seekers**. Migration policies have always been a source of disagreement and debate in Australia. The implications of the scale of Australia's current rates of immigration on Australia's major cities are discussed in subtopic 3.7.

3.4.4 Summary of factors affecting population change in Australia

There has been a range of social, economic, social and political factors involved in the changes in Australia's population over time. These factors include the changing role of women in society, improvements in health care, the availability of family planning techniques, the provision of government incentives such as baby bonuses, farm consolidation and the rate of urbanisation, and migration policies. Over time, factors such as these have led to changes in Australia's population such as falling birth and death rates, increasing life expectancy and increasing rates of migration to Australia.

Activity 3.4: Analyse and explain factors involved in changes to Australia's population

Explain and analyse the data

1. Refer to figures 3.17 and 3.19. Explain why life expectancy gains in Australia levelled out, especially for males, in the period from the 1940s to the 1970s, before large gains in life expectancy were made from the 1970s.
2. Refer to the list of social, economic and cultural factors that have had an impact on birth rates in Australia in section 3.4.2. Explain why each of these factors might have had an impact on Australia's birth rates.

Synthesise and apply your knowledge

3. Identify possible links between the factors affecting population change in Australia by mind-mapping the interrelationships between factors.

4. Use table 3.5 to complete this question. Possible techniques to use for this task are scattergraphs and Spearman's rank correlation co-efficient.
   (a) Choose a graphical or mathematical technique for examining the connection between changes in female workforce participation rates, birth rates and fertility rates over the period 1960–2018, and then draw the graphs or complete the calculations.
   (b) Explain what your chosen technique shows about the connection between female workforce participation rates and birth and fertility rates.

Research and explain push and pull factors

5. Suggest possible pull factors that have been responsible for Queensland's attraction for inter-state migrants. What might be some of the push factors for New South Wales?
6. The Australian government's migration policies, especially regarding the number of migrants accepted, often change year to year. Research the government's current migration policies, the components of the migration program and the numbers of people accepted or planned for under the program's various components.

CALCULATING CORRELATIONS: SPEARMAN'S RANK CORRELATION COEFFICIENT

The scattergraph and line of best fit are useful for identifying any anomalies (outliers or residuals) that don't fit in with a general pattern of results. An alternative method is to use a statistical test. This establishes whether the correlation is statistically significant or if it could have been the result of chance alone. Spearman's rank correlation coefficient is a technique that can only yield a result between 1 and minus 1. You must be comparing at least five pairs of data for the test to work.

METHOD — CALCULATING THE COEFFICIENT

Step 1 Lay out your data in a table.
Step 2 Rank the two data sets. To do this, assign '1' to the largest number, '2' to the second largest and so on. The smallest value in the column will be assigned the lowest ranking. Do this for both sets of measurements. If two or more values are the same, or 'tied', they are assigned the average rank. For example, if there are three tied scores that are second they will all be ranked 3, and then the ranks will continue (i.e. 1, 3, 3, 3 , 5, 6 and so on). If there are only two tied ranks, say 7, they become 7.5, 7.5 and then 9, 10 and so on.
Step 3 Find the difference in the ranks (d). This is the difference between the ranks of the two values on each row of the table. Do not worry about the signs +ve or –ve because they are squared in the next step.
Step 4 Square each rank difference you have just calculated.
Step 5 Add up all these squared differences.
Step 6 Substitute your answers into the formula. The formula has two constant values, 1 and 6.
- Σ = sigma (sum of)
- d = difference between ranks
- n = number of pairs in your data (the number of pairs in your table)

$$r_s = 1 - \frac{6 \sum d^2}{n^3 - n}$$

For example, to determine if there is a correlation between the size of a town and the number of petrol stations, you might consider the following data:

Population	Rank (pop.)	Number of petrol stations	Rank (no. of stations)	Difference between ranks	Difference between ranks squared
30 000	4	6	5	1	1
20 000	5	5	4	1	1
45 000	2	8	2	0	0
39 000	3	7	3	0	0
58 000	1	9	1	0	0
					$\sum d_2 = 2$

$$1 - \left(\frac{6 \times 2}{5^3 - 5}\right)$$

$$1 - \frac{12}{120}$$

$$1 - 0.1 = +0.9$$

The closer the value is to +1 or −1, the stronger the likely correlation (i.e. the likelihood that one variable causes another). A perfect positive correlation is +1 and a perfect negative correlation is −1. This value of +0.9 suggests a very strong positive relationship.

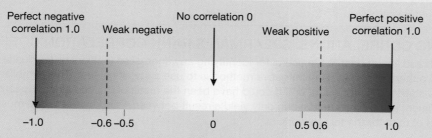

FIGURE 3.26 Positive and negative correlation scale

A Spearman's rank significance graph can be used to test the significance of the relationship by looking up the value of +0.9.

Begin your test by working out the 'degrees of freedom'. This is the number of pairs in your sample minus 2 (n–2). In the example, it is 3 (5 − 2 = 3 degrees).
Plot your result on the Spearman's rank significance graph. The x-axis shows the Spearman's rank correlation coefficient and the y-axis shows the degrees of freedom.

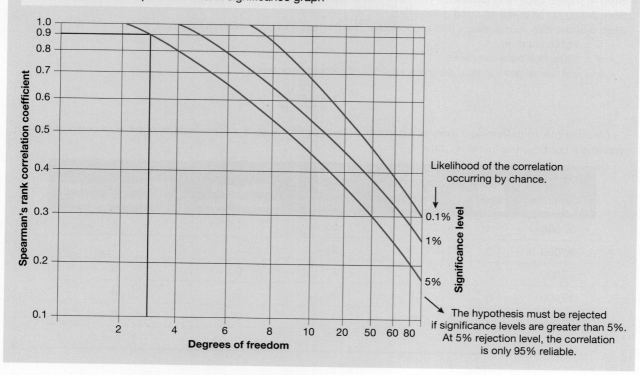

FIGURE 3.27 The Spearman's rank significance graph

Although +0.9 is a high coefficient, our result is below the 5 per cent significance level, therefore it is quite possible the result was the product of chance and statistically that is not considered significant. The sample of 5 towns is, after all, exceedingly small. If the result had been above the 0.1 per cent significance level, then we could be 99.9 per cent confident the correlation had not occurred by chance.

Remember that a correlation between two sets of variables does not prove that one variable causes another variable. It can suggest there is a relationship but only further research can actually prove that one thing affects the other. The size of the sample is important. The more data you have, the more reliable your result.

on Resources

Video eLessons SkillBuilder: Constructing a box scattergram (eles-1734)

SkillBuilder: Constructing and interpreting a scattergraph (eles-1756)

Interactivities SkillBuilder: Constructing a box scattergram (int-3352)

SkillBuilder: Constructing and interpreting a scattergraph (int-3374)

3.5 Population patterns and trends

3.5.1 Population distribution

There are several key features of the distribution of Australia's population.
1. There is a high degree of geographical concentration — a large percentage of Australians live in the south-east and in a smaller part of the south-west of the continent.
2. The population is further concentrated in the largest cities. Around 71 per cent of Australians live in major cities, and two-thirds live in the capital cities. In total, approximately 88 per cent of Australians live in urban settlements of some kind.
3. The population is predominantly located on or near the coast — approximately 85 per cent of Australians live within 50 km of the coast.
4. A significant area of Australia is very sparsely populated. Only Victoria has almost nowhere with very low population density, while significant areas of Western Australia, South Australia, the Northern Territory and Queensland are very sparsely populated.

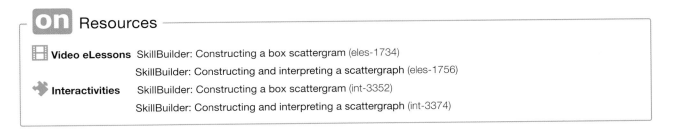

FIGURE 3.28 Australia's uneven population distribution

There are both geographical and historical reasons for the distribution pattern of Australia's population. Australia's aridity explains very low population densities across much of the continent, while higher rainfall is partly responsible for higher population densities in coastal areas. The capital cities were often the points of first settlement in each state and became the major ports for the export of primary products, as well as the administrative and financial centres of the state. This initial advantage for the capital cities had a **multiplier effect** on population growth, and has led to their dominance in Australia's settlement hierarchy.

The population numbers for each of Australia's states and territories reflect the distribution patterns. Figure 3.29 shows the population by state and territory from the 2016 Census of Population and Housing. The majority of people live in the eastern mainland states, with New South Wales the most populous state, followed by Victoria and Queensland. Almost 80 per cent of Australians live in these three states and the Australian Capital Territory.

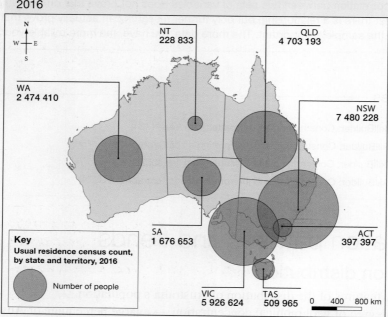

FIGURE 3.29 Usual residence census count, by state and territory, 2016

NT 228 833
QLD 4 703 193
WA 2 474 410
NSW 7 480 228
SA 1 676 653
ACT 397 397
VIC 5 926 624
TAS 509 965

Key
Usual residence census count, by state and territory, 2016
Number of people

Source: Australian Bureau of Statistics

Activity 3.5a: Patterns of population density, birth and death rates, and rates of natural increase

For this activity, you will need to use table 3.2 from activity 3.2c on page 126.

Analyse population density data and apply your knowledge

1. Use the population density figures to describe Australia's population density by state and territory.
2. Compare birth and death rates and rates of natural increase for the states and territories.
3. Construct a map, similar to figure 3.29, to illustrate population density by state and territory in 2017. Figure 3.29 has used proportional circles to illustrate the relative populations of the states and territories. Your map should use a similar method to show population density. Use **SkillBuilder: Constructing and describing proportional circles on maps** in the Resources tab for assistance.

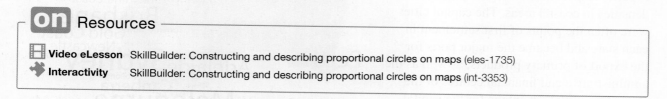

Resources

Video eLesson   SkillBuilder: Constructing and describing proportional circles on maps (eles-1735)
Interactivity   SkillBuilder: Constructing and describing proportional circles on maps (int-3353)

3.5.2 Age–sex patterns

Age affects where people live. Major urban areas have a higher proportion of younger people and a lower proportion of people aged 65 years and above than other parts of Australia, where there are more older people than younger people. Education and work opportunities in capital cities are pull factors for younger adults, but older people often move away from busy urban areas when they retire, and are also more likely to work in rural areas.

Figure 3.30 illustrates the age structure of the capital cities and other major urban areas compared with smaller towns, villages and rural areas. The capital cities have a much greater percentage of people in the 20 to 44 age cohorts than the rest of Australia (38 per cent of the capital cities' populations compared with 30 per cent for the rest of Australia). In contrast, the rest of Australia had a greater percentage of people in the older age cohorts (45 per cent aged over 45) than the capital cities (37 per cent).

In terms of the sex structure of the population, there is little difference between the capital cities and the rest of Australia. Females outnumber males in both capital cities (8.35 million females to 8.21 million males) and the remainder of the country (4.05 million females to 4.00 million males). This difference is largely a consequence of the greater life expectancy of females.

The age structure of the population differs between states and territories as well as between major urban areas and other parts of Australia. In 2016, Tasmania and South Australia had the highest proportion of their populations aged 65 years and above, while the Northern Territory had the lowest.

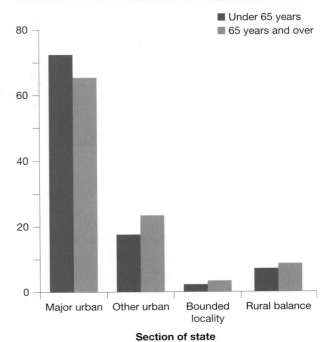

FIGURE 3.30 Proportion of people aged 65 and above by section of state, 2016

Source: Australian Bureau of Statistics

FIGURE 3.31 Age and sex distribution (%), greater capital cities and rest of Australia, to 30 June 2017

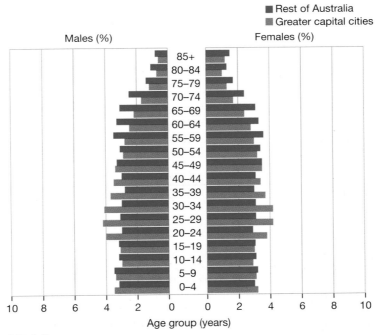

Source: Australian Bureau of Statistics

FIGURE 3.32 Proportion of people aged 65 and above by state and territory, 2006 and 2016

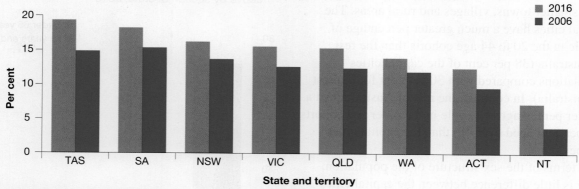

Source: Australian Bureau of Statistics

The median age (the age at which half the population is younger and half is older) of the populations of the states and territories also illustrates this age pattern. The median age in Australia in 2017 was 37.3 years — for males it was 36.4 years and for females it was 38.1 years. Table 3.7 shows the median ages of the population in each state or territory and in their capital cities in 2017. The Northern Territory had the lowest median age, while Tasmania had the highest. Their capital cities also reflect this pattern, with Darwin having the lowest and Hobart the highest median ages.

TABLE 3.7 Median ages by state or territory and Greater Capital City Statistical Areas (GCCSAs), 2017

State or territory	Median age (years)	Population aged 65 years and over (%)	GCCSA	Median age (years)
NSW	37.5		Greater Sydney	35.8
VIC	36.8		Greater Melbourne	35.6
QLD	37.1		Greater Brisbane	35.3
SA	40.0		Greater Adelaide	38.7
WA	36.6		Greater Perth	36.1
TAS	42.2		Greater Hobart	39.8
NT	32.6		Greater Darwin	33.6
ACT	35.0		Australian Capital Territory	35.0

Source: Australian Bureau of Statistics

Activity 3.5b: Age–sex patterns
Explain and analyse data

1. Use figure 3.32 to complete the 'Population aged 65 years and over' column in table 3.7.
2. Given the data provided in table 3.7, explain why the patterns evident in figure 3.30 would probably not be the case in the Northern Territory.
3. Refer to table 3.8. Describe the regional patterns of population in Queensland for the three demographic indicators.
4. Choose one of the indicators. Construct a choropleth map to illustrate Queensland's regional patterns for that demographic indicator. You can use a copy of figure 3.33 as the base map for this task.

TABLE 3.8 Queensland demographic statistics, 2016

Statistical area	Population (persons)	Population density (persons per km^2)	Median age (years)
Brisbane — East	223 095	341.6	40
Brisbane — North	206 522	1 104.7	37
Brisbane — South	340 569	1 283.5	34
Brisbane — West	178 991	663.8	36
Brisbane — Inner City	250 207	3 061.0	33
Cairns	240 190	11.3	39
Central Queensland	220 912	1.9	36
Darling Downs–Maranoa	126 289	0.8	41
Gold Coast	569 997	306.8	39
Ipswich	323 069	48.4	34
Logan–Beaudesert	317 296	122.7	34
Mackay–Isaac–Whitsunday	169 688	1.9	37
Moreton Bay — North	236 091	54.3	41
Moreton Bay — South	194 969	252.1	35
Queensland — Outback	79 700	0.1	33
Sunshine Coast	346 522	112.3	44
Toowoomba	149 512	66.2	37
Townsville	229 031	2.9	36
Wide Bay	287 883	5.9	46
Queensland	**4 703 193**	**2.7**	**37**

Source: © The State of Queensland Queensland Treasury 2017

FIGURE 3.33 Queensland's statistical areas

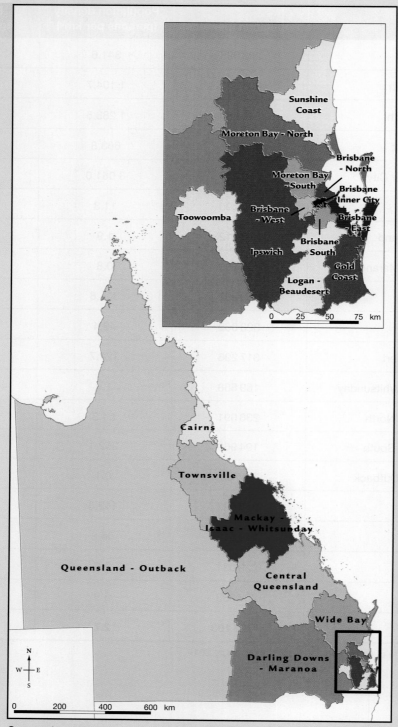

Source: Australian Bureau of Statistics, MapData Services Pty Ltd, PSMA

on Resources

- **Digital document** Queensland statistical areas outline map (doc-31989)
- **Video eLessons** SkillBuilder: Reading and describing basic choropleth maps (eles-1706)
  SkillBuilder: Constructing and describing complex choropleth maps (eles-1732)
  SkillBuilder: Using Excel to construct population profiles (eles-1758)
- **Interactivities** SkillBuilder: Reading and describing basic choropleth maps (int-3286)
  SkillBuilder: Constructing and describing complex choropleth maps (int-3350)
  SkillBuilder: Using Excel to construct population profiles (int-3376)

3.5.3 Migrant settlement patterns

The migrant population throughout Australia is increasing. New South Wales and Victoria, Australia's most populous states, have the highest proportions of overseas-born people. In 2016, nearly 61 per cent of migrants lived in these two states, but increasing numbers are also settling in Queensland and Western Australia. In 1966, the proportion of overseas-born people living in these two states was 9.5 per cent and 9.3 per cent respectively, compared with 16.5 per cent and 12.9 per cent in 2016. The Northern Territory, with 0.7 per cent, and Tasmania, with 1 per cent, have relatively low numbers of overseas-born people.

FIGURE 3.34 Overseas-born population of states and territories, 2016

NT 45 403 0.7%
QLD 1 015 875 16.5%
WA 797 695 12.9%
SA 384 097 6.2%
NSW 2 072 454 33.6%
VIC 1 680 256 27.3%
ACT 105 161 1.7%
TAS 61 240 1.0%

Key
Overseas-born population of states and territories, 2016
NT — State
45 403 — Number of people
0.7% — Percentage

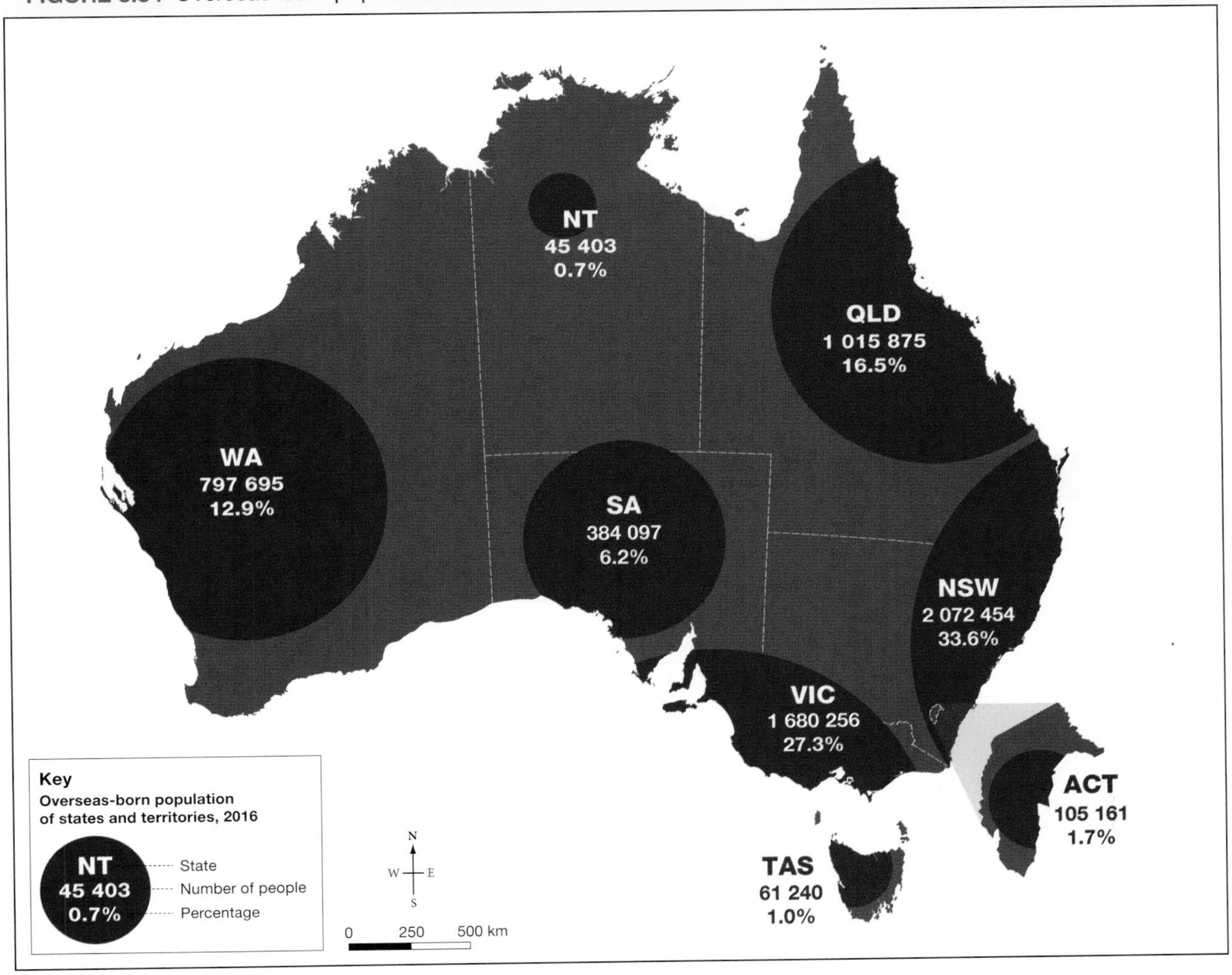

Source: Australian Bureau of Statistics

Migrants are concentrated in the capital cities. Sydney had the largest overseas-born population in 2016 at 1 773 496 people, nearly 40 per cent of Sydney's total population. Next were Melbourne, with 1 520 253 overseas-born people (36 per cent of the population), and Perth, with 702 545 (39 per cent of the population). In general, people born overseas are more likely to live in a capital city than Australian-born people. In 2016, 83 per cent of overseas-born people lived in capital cities, compared with only 61 per cent for Australian-born people.

The distribution of migrants across the capital cities does vary when countries of origin are considered. In Sydney, Chinese-born migrants were the largest overseas-born group in 2016, representing 5 per cent of Sydney's population. In Melbourne, Indian-born migrants were the largest group, while in Brisbane, New Zealand-born migrants were the largest and migrants to Darwin were most commonly born in the Philippines. Table 3.9 lists the top three countries of origin for each of the capital cities at the time of the 2016 census.

TABLE 3.9 Birth places and percentage of population of overseas-born residents, 2016

Sydney	%	Melbourne	%	Brisbane	%	Perth	%
China	5.0	India	3.8	New Zealand	5.0	England	9.2
England	3.4	China	3.7	England	4.2	New Zealand	3.4
India	2.9	England	3.2	China	1.7	India	2.6
Adelaide	**%**	**Hobart**	**%**	**Darwin**	**%**	**Canberra**	**%**
England	6.6	England	3.8	Philippines	4.1	England	4.0
India	2.1	China	1.1	England	3.5	China	3.0
China	2.0	New Zealand	0.9	New Zealand	2.4	India	2.8

Source: George Megalogenis / Australian Foreign Affairs

Table 3.10 provides data on migrant settlement patterns in Queensland. Brisbane has the highest number of overseas-born people, followed by the Gold Coast, but all parts of Queensland have a migrant population. The greatest concentration of migrants occurred in the south Brisbane area, where 36.9 per cent of the total population in 2016 was born overseas. Outback Queensland, where there are far fewer work and education opportunities, had the lowest number of migrants.

TABLE 3.10 Queensland's overseas-born population by statistical area, 2016

Statistical area	Persons	% of total persons
Brisbane — East	50 916	22.8
Brisbane — North	49 984	24.2
Brisbane — South	125 675	36.9
Brisbane — West	52 226	29.2
Brisbane — Inner City	71 833	28.7
Cairns	45 928	19.1
Central Queensland	24 294	11.0
Darling Downs–Maranoa	10 265	8.1
Gold Coast	160 312	28.1
Ipswich	73 205	22.7

(Continued)

TABLE 3.10 Queensland's overseas-born population by statistical area, 2016 (Continued)

Statistical area	Persons	% of total persons
Logan–Beaudesert	84 415	26.6
Mackay–Isaac–Whitsunday	20 759	12.2
Moreton Bay — North	43 143	18.3
Moreton Bay — South	41 203	21.1
Queensland — Outback	7204	9.0
Sunshine Coast	68 824	19.9
Toowoomba	19 190	12.8
Townsville	28 766	12.6
Wide Bay	34 557	12.0
Queensland	**1 015 875**	**21.6**

Source: © The State of Queensland Queensland Treasury 2017 based on ABS data

Activity 3.5c: Migrant settlement patterns

Apply and analyse the local population data

1. Fill table 3.11 using data from figures 3.29 and 3.34. Use the data to calculate the percentage of the total population of each state and territory made up of people born overseas.

TABLE 3.11 Percentage of population overseas-born, state or territory, 2016

State or territory	Total population	Overseas-born population	Proportion of population overseas-born (% of state/territory population)
NSW			
VIC			
QLD			
SA			
WA			
TAS			
NT			
ACT			
Australia			

Describe and compare patterns of data

2. Describe the pattern of the overseas-born population illustrated by your calculations.
3. Using table 3.9, compare the places of origin for overseas-born people in Australia's capital cities. Look for similarities and differences in the patterns.
4. Refer to the website **SBS immigration facts** in the Resources tab, which provides Australia-wide data on where Australia's migrants were born. Describe the overall pattern of migrant settlement in Queensland. In particular, look for the regions where there were no migrants.

5. Locate the area in which you live on the map and list the main countries of origin of migrants including and excluding migrants from England and New Zealand.
6. Compare your local area with two or three other locations in Queensland.

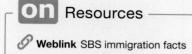

Weblink SBS immigration facts

3.5.4 Indigenous Australians

Aboriginal and Torres Strait Islander people live across the country, but have larger populations in some states than others (see table 3.12). In 2016, the largest number of Indigenous Australians lived in New South Wales, with Queensland second. In total, 62 per cent of Indigenous people lived in these two states. In each of these states, Indigenous Australians represent a relatively small percentage of the total population, much smaller than the percentage of overseas-born people. However, Indigenous Australians make up over a quarter of the population of the Northern Territory, and also outnumber overseas-born people.

In 2016, 35 per cent of Indigenous Australians lived in capital cities, but there were wide variations between the states and territories. In the Northern Territory, 78 per cent of Indigenous Australians lived outside Darwin, and in Queensland, 71 per cent lived outside Brisbane. In contrast, 50 per cent or more of South Australia's and Victoria's Indigenous people lived in Adelaide and Melbourne.

TABLE 3.12 Aboriginal and Torres Strait Islander people by state and territory, 2016

	Total Aboriginal and Torres Strait Islander people		As a proportion of state/territory population (%)
	Number	Per cent	
NSW	216 176	33.3	2.9
VIC	47 788	7.4	0.8
QLD	186 482	28.7	4.0
SA	34 184	5.3	2.0
WA	75 978	11.7	3.1
TAS	23 572	3.6	4.6
NT	58 248	9.0	25.5
ACT	6508	1.0	1.6
Australia	649 171	100.0	2.8

Source: Australian Bureau of Statistics

In comparison to the rest of Australia's population, Indigenous Australians are relatively young, with a median age of 23, and 34 per cent of the population aged below 15 years. One result of their younger overall age is that the fertility rate for Indigenous Australians is higher than for Australia as a whole (2.3 children per Indigenous woman compared with 1.8 per non-Indigenous woman).

FIGURE 3.35 Population pyramids for Indigenous and non-indigenous Australians

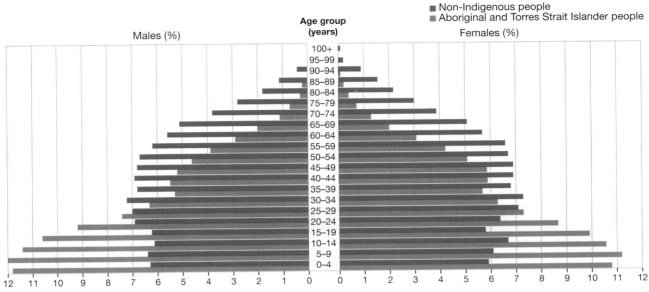

Source: Australian Bureau of Statistics

Despite the age difference between Indigenous and non-Indigenous Australians, life expectancy and death rates for Indigenous Australians are worse than for non-Indigenous Australians. In 2016, life expectancy for Indigenous males was 71.6 years and 75.6 years for Indigenous females compared with 80.2 and 83.4 for Australians overall. The death rate for Indigenous Australians was 9.8 per 1000 compared with 5.9 for Australia as a whole. It is clear that Aboriginal and Torres Strait Islander peoples continue to suffer social and economic disadvantage, particularly in terms of health care. Table 3.13 provides you with some indication of the economic position of Indigenous Australians relative to Australia as a whole.

TABLE 3.13 Economic data for Indigenous Australians, 2016

	Indigenous Australians	Australia overall
Unemployment (%)	18.2	6.9
Median personal weekly income ($)	441	662
Median household weekly income ($)	1203	1438

Source: Australian Bureau of Statistics

Activity 3.5d: Analysing Indigenous population data

Explain and analyse the data

1. Compare the differences in the proportion of population aged under 15 years and 65 years and above for Indigenous and non-Indigenous Australians. Reflect on the possible reasons for these differences.
2. What might be some of the implications of the age structure of Australia's Indigenous population for Aboriginal and Torres Strait Islander people? For Australia as a whole?
3. What might be done to overcome some of the social, health and economic challenges faced by Indigenous Australians?

3.6 Local area population patterns

The demographic characteristics of Australia's states and of regions within states can vary a great deal. The differences in the demographic character of smaller local areas such as regional cities, country towns and the suburbs of cities can vary even more.

This subtopic provides you with data on several suburbs and country towns, so you can identify, analyse and compare demographic patterns across a variety of local areas, particularly in Queensland. An initial comparison could be made with the national and state-wide data shown in table 3.14, before considering similarities and differences in the demographic characteristics of the suburbs and country towns. You are also encouraged to obtain the Australian Bureau of Statistics data for your local area to compare with the places provided in the text.

TABLE 3.14 Demographic statistics for Queensland and Australia, 2016

Demographic characteristic	Queensland	Australia
Population	4 703 193	23 401 892
Male population (%)	49.4	49.3
Female population (%)	50.6	50.7
Population 0–14 years (%)	19.4	18.7
Population 65 years and above (%)	15.2	18.5
Median age	37	38
Average number of children per family		
for families with children	1.9	1.8
for all families	0.7	0.8
Aboriginal and Torres Strait Islander population	186 482 (4.0%)	649 171 (2.8%)
People born overseas	1 359 536 (28.9%)	7 787 051 (33.3%)
Country of birth (% of total population)	Australia 71.1	Australia 66.7%
	China 1.0	England 3.9
	Taiwan 0.3	New Zealand 2.2%

Source: Australian Bureau of Statistics

FINDING AUSTRALIAN BUREAU OF STATISTICS DEMOGRAPHIC DATA

The Australian Bureau of Statistics (ABS) conducts a census of Australia's population every five years. The data collected during the census includes information on age, sex, marital status, family composition, level of education, country of birth, occupation and income. ABS data for the most recent census (2016) is available on the website in the Resources tab. To find local data, you should enter the name of your local area in the QuickStats search.

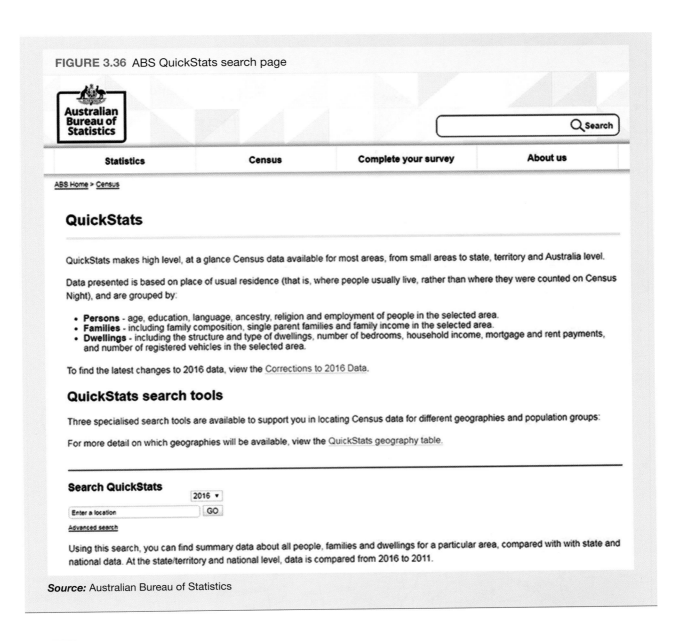

FIGURE 3.36 ABS QuickStats search page

Source: Australian Bureau of Statistics

Resources

Weblink Australian Bureau of Statistics

3.6.1 Comparing demographic patterns of suburbs

Boondall and Sunnybank are middle distance residential suburbs of Brisbane. Boondall is located approximately 15 km north-east of the CBD and Sunnybank approximately 12 km south of the CBD. Boondall is a long-established residential suburb, dating back to the late 1800s, while Sunnybank's residential housing development began in the 1950s.

FIGURE 3.37 Location map: Boondall and Sunnybank

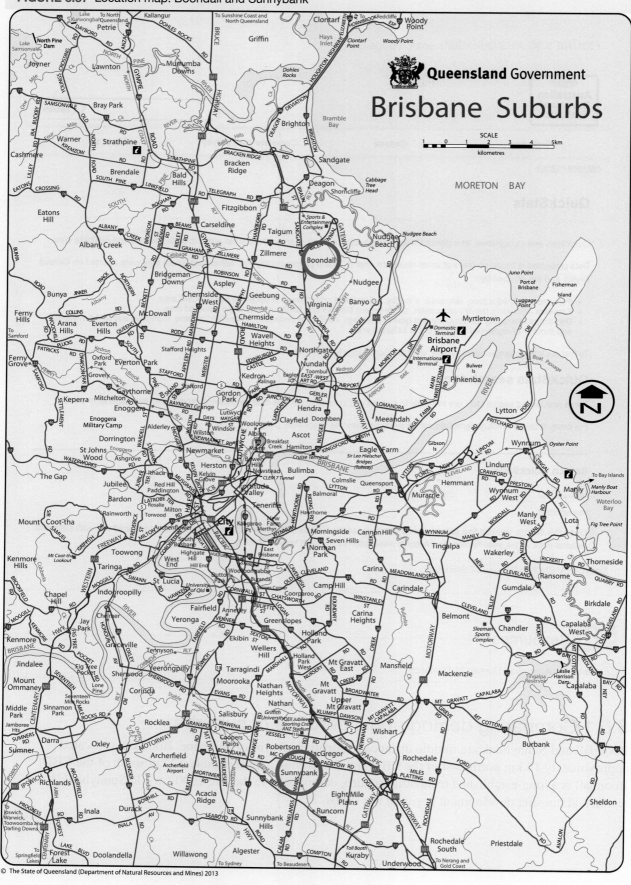

FIGURE 3.38 Aerial photo of Boondall

Source: Oz Aerial Photography

FIGURE 3.39 Aerial photo of Sunnybank

Source: © Mike Swaine

The residential suburb of Springfield is located approximately 20 km south-east of the City of Ipswich and 33 km south-west of Brisbane's CBD. Springfield is part of a master-planned urban development begun in the 1990s by the Springfield City Group. It is one of six suburbs that make up Greater Springfield, the first privately built city in Australia.

FIGURE 3.40 Aerial photo of Springfield

CHAPTER 3 Population challenges in Australia **161**

FIGURE 3.41 Location map: Springfield

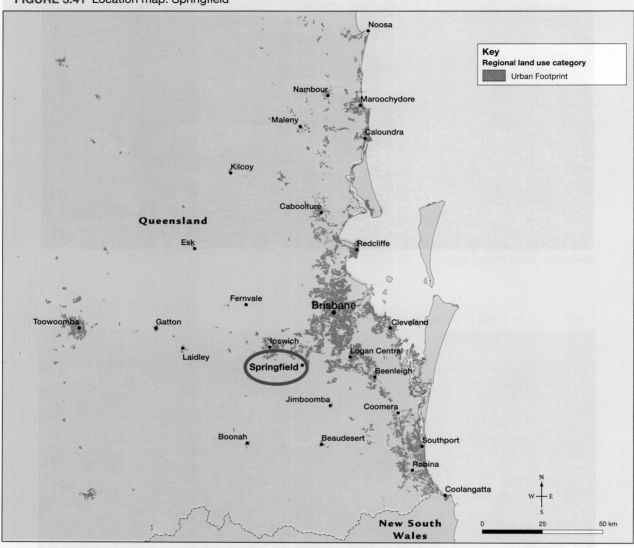

Source: The State of Queensland, Department of Infrastructure, Local Government and Planning

Harris Park, in New South Wales, is a residential suburb of the City of Parramatta and part of Greater Western Sydney. It is located 23 km west of the Sydney CBD. Residential development began in the 1880s with the opening of a railway station (figure 3.42). From the 1960s, development increasingly involved the construction of home units rather than separate houses. Parramatta was designated a major growth centre after World War II and saw a large increase in the number of overseas migrants living in Harris Park.

FIGURE 3.42 Harris Park railway station

FIGURE 3.43 Location map: Harris Park

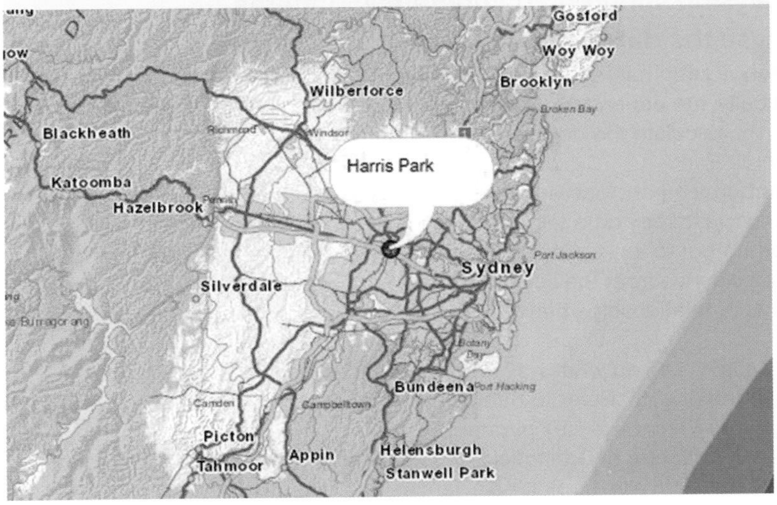

Source: Geoscience and Natural Earth

TABLE 3.15 Demographic statistics for Boondall, Sunnybank, Springfield and Harris Park, 2016

Demographic characteristic	Boondall	Sunnybank	Springfield	Harris Park
Population	9217	8697	6722	5799
Male population (%)	50.6	50.3	49.6	53.6
Female population (%)	49.4	49.7	50.4	46.4
Population 0–14 years (%)	17.9	12.7	27.0	17.7
Population 65 years and above (%)	16.5	14.3	5.4	7.0
Median age	37	32	30	31
Average number of children per family				
for families with children	1.8	1.8	2.0	1.4
for all families	0.7	0.6	1.2	0.6
Aboriginal and Torres Strait Islander population	231 (2.5%)	84 (1%)	210 (3.1%)	19 (0.3%)
People born overseas	2914 (31.3%)	5456 (62.7%)	2321 (33.8%)	4706 (81.1%)
Country of birth (% of total population)	Australia 68.7	Australia 37.3	Australia 66.2	Australia 18.9
	New Zealand 5.3	China 19.0	New Zealand 10.2	India 46.4
	India 4.1	Taiwan 9.3	England 6.8	China 5.1

Source: Australian Bureau of Statistics

Activity 3.6a: Analysing suburban demographic patterns

Explain and analyse the data

1. The term **dependency ratio** refers to the percentage of a population dependent on people who are in the workforce. Statistically, the dependent population is those aged below 15 and above 65 years of age. Using tables 3.14 and 3.15, calculate the dependency ratios for Australia, Queensland, Boondall, Sunnybank, Springfield and Harris Park.
   (a) Which of the suburbs most closely reflects Australia's and Queensland's dependency ratios?
   (b) What does the dependency ratio tell us about the age structure of each of the suburbs?
   (c) Why is it important not to rely solely on the dependency ratio when analysing age-related characteristics of places? Use evidence from the suburbs in table 3.15 to support your answer.
2. Identify connections or relationships between two or more of the demographic characteristics provided in table 3.15.
3. Refer to figures 3.44 and 3.45. In what ways do the age–sex pyramids for Springfield and Harris Park reflect the data provided in table 3.15? Refer to particular age cohorts in your answer. What might explain the peculiar shape of Harris Park's age–sex pyramid?
4. Compare the age–sex structure of Springfield and Harris Park with that of their associated cities (Ipswich and Greater Sydney respectively). Look for similarities and differences.
5. Which, if any, of the suburbs most closely resembles the demographic characteristics of Australia and Queensland? If there is a suburb, describe how it resembles Australia/Queensland.
6. Calculate location quotients for Australian-born residents in each of the suburbs (see box that follows). What do your results tell you about the concentration of Australian-born residents in each of these suburbs?
7. Select one of the suburbs and write a paragraph or two in which you describe its demographic features. What possible economic, social and environmental challenges might the suburb face as a consequence of these demographic features? What actions might be taken to manage these challenges? Will your proposed actions be economically, socially and environmentally sustainable?

FIGURE 3.44 Age–sex pyramids for Springfield and Ipswich City, 2016

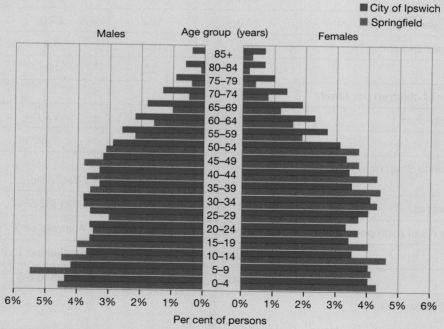

Source: © 2010–2019 .id consulting pty ltd

FIGURE 3.45 Age–sex pyramids for Harris Park and Greater Sydney, 2016

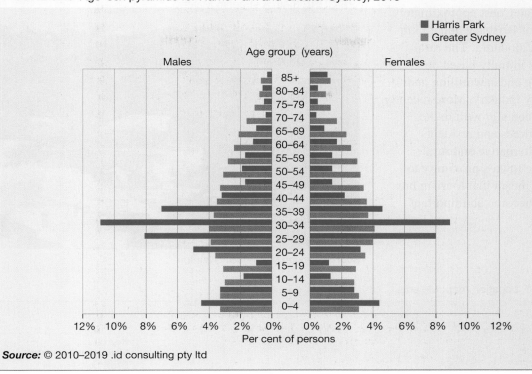

Source: © 2010–2019 .id consulting pty ltd

LOCATION QUOTIENTS

A **location quotient** (LQ) is a way of quantifying how concentrated a demographic feature is in a particular region compared to the country as a whole. It is calculated using the formula:

$$LQ = \frac{S/\sum(S)}{S/\sum(T)}$$

$S/\sum(S)$ is the concentration of the demographic feature in the region. $S/\sum(T)$ is the concentration in the country as a whole.

S is the total for the demographic feature, while $\sum(S)$ and $\sum(T)$ are the overall total for the region and country respectively. Often the concentration for both region and country is given as a percentage. For example, in Brisbane in 2016 the percentage of people born in Australia was 67.8 (S = 67.8, $\sum(S)$ = 100), while for Sydney it was 57.1 (S = 57.1, $\sum(S)$ = 100).

The location quotients for the Australian-born population in Brisbane and Sydney can now be calculated.

$$\text{Brisbane: } \frac{67.8/100}{66.7/100}$$

$$LQ = 1.01$$

$$\text{Sydney: } \frac{57.1/100}{66.7/100}$$

$$LQ = 0.86$$

If the LQ for the demographic feature being analysed is greater than 1.0, a higher concentration of that feature is found in the region, while a LQ below 1.0 means a lower concentration. In general, a LQ of 1.2 or higher indicates some degree of concentration of the demographic feature being analysed, while a LQ below 0.8 indicates a low degree of concentration. In this case both Brisbane and Sydney fall within the 'normal' distribution for Australian-born residents.

3.6.2 Comparing demographic patterns of country towns

Maleny is located in the hinterland of the Sunshine Coast approximately 80 km north of Brisbane and 25 km inland from Caloundra. The rural township was initially based on timber-cutting and sawmilling and, later, the dairy industry. More recently, Maleny has seen a growth in the number of retirees and residents looking for alternative and rural lifestyles. The town's proximity to Brisbane has meant that tourism has increasingly become an important industry.

FIGURE 3.46 A street in Maleny

Source: Lyndon Mechielsen / Newspix

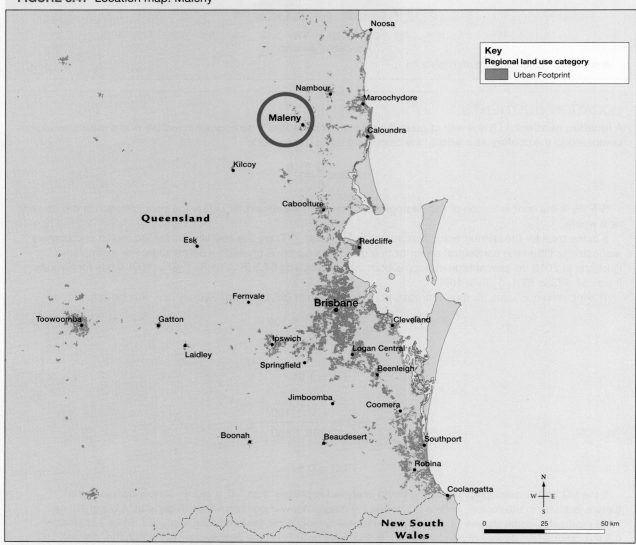

FIGURE 3.47 Location map: Maleny

Source: The State of Queensland, Department of Infrastructure, Local Government and Planning

166 Jacaranda Senior Geography for Queensland 2 Units 3 & 4 Third Edition

FIGURE 3.48 A view of Cloncurry

Source: © 2019 Aussie Towns

Cloncurry is a rural town located in north-western Queensland approximately 800 km west of Townsville and 1700 km north-west of Brisbane. Cloncurry Township was surveyed in 1876 and the first general store was established in the same year. Initial European settlement in the region was based on the pastoral industry (related to livestock, usually sheep or cattle). The discovery of copper and gold in the 1880s resulted in the growth of the mining industry near the town. The arrival of the railway in 1907 was another stimulus for Cloncurry's growth. The town also became the headquarters of the Australian Inland Mission's flying doctor service. There has been some pressure on employment and services in recent years with the downturn in mining.

FIGURE 3.49 Location map: Cloncurry

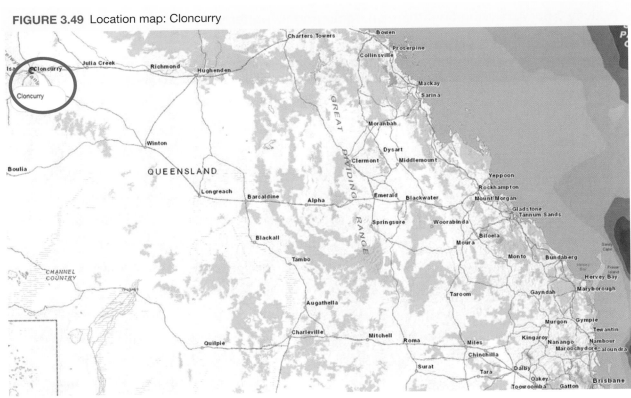

Source: Geoscience and Natural Earth

TABLE 3.16 Demographic statistics for Maleny and Cloncurry, 2016

Demographic characteristic	Maleny	Cloncurry
Population	3743	2254
Male population (%)	47	52.5
Female population (%)	53	47.5
Population 0–14 years (%)	13.4	22.0
Population 65 years and above (%)	36.3	9.0
Median age	57	34
Average number of children per family		
for families with children	1.7	1.9
for all families	0.5	0.7
Aboriginal and Torres Strait Islander population	49 (1.3%)	533 (23.7%)
People born overseas	980 (25.5%)	570 (24.6%)
Country of birth (% of total population)	Australia 74.5	Australia 75.4
	England 7.9	New Zealand 2.5
	New Zealand 3.3	England 1.0
		Philippines 1.0

Source: Australian Bureau of Statistics

Activity 3.6b: Analysing country town demographic patterns

TABLE 3.17 Cloncurry's population by age, 2016

Age	Number of people	Percentage of population
0–4 years	197	8.8
5–9 years	160	7.1
10–14 years	136	6.1
15–19 years	123	5.5
20–24 years	142	6.3
25–29 years	171	7.6
30–34 years	219	9.8
35–39 years	165	7.3
40–44 years	123	5.5
45–49 years	152	6.8
50–54 years	152	6.8

(continued)

TABLE 3.17 Cloncurry's population by age, 2016 *(continued)*

Age	Number of people	Percentage of population
55–59 years	167	7.4
60–64 years	137	6.1
65–69 years	78	3.5
70–74 years	65	2.9
75–79 years	22	1.0
80–84 years	19	0.8
85+ years	17	0.8

Source: Australian Bureau of Statistics

1. Refer to tables 3.16 and 3.17 and figure 3.50. Comment on the age–sex characteristics of Maleny and Cloncurry. What might account for the differences between the two?
2. Compare Maleny's age–sex pyramid with that of Greater Brisbane. What does Maleny's age structure tell us about Maleny's population? What are the possible implications of this age structure for Maleny?
3. Compare Cloncurry's age structure (table 3.17) with that of rural North Queensland (figure 3.51). In what ways is Cloncurry's age structure similar to or different from North Queensland?
4. What possible challenges might Cloncurry face as a result of its demographic characteristics? What actions might be taken to manage these challenges? What could Cloncurry do to reverse its population decline and help make the town more socially and economically sustainable?

FIGURE 3.50 Age–sex pyramids for Maleny and Greater Brisbane, 2016

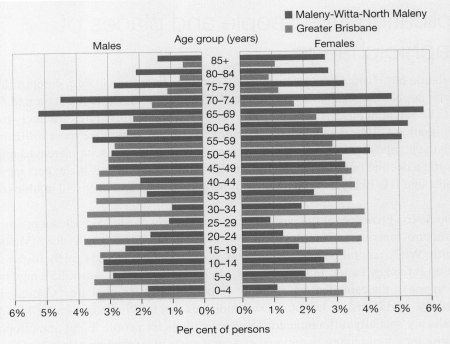

Source: © 2010–2019 .id consulting pty ltd

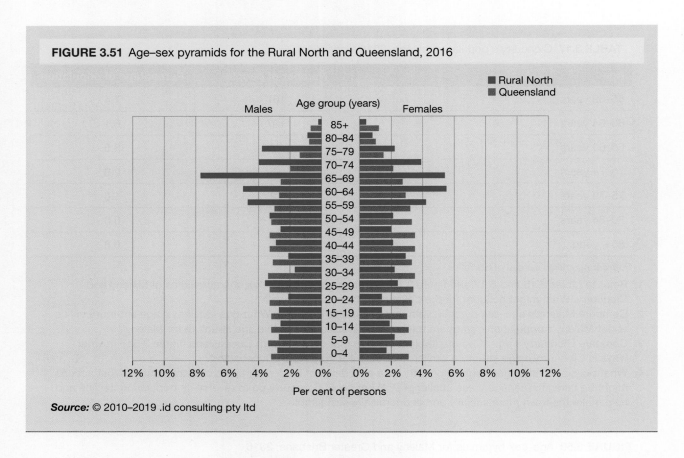

FIGURE 3.51 Age–sex pyramids for the Rural North and Queensland, 2016

3.7 Implications for people and places of demographic change

Australia's population has changed over time and will continue to change in size, composition and age structure. The total population reached 25 million in 2018 and is predicted to reach around 40 million by the mid-2050s. (The actual number is difficult to forecast — while fertility rates are likely to remain around 1.9, rates of immigration will be dependent on government policies, which are more difficult to predict). Current trends in the composition of the population are likely to continue, with increased migration from Asian countries and a decline in European migration. It is almost certain that life expectancy will continue to rise in Australia, and it is likely that the number of people aged 65 and over will double over the next 40 years.

Australia's population growth over the past 20 years has already created significant pressure on services, infrastructure, housing and employment, especially in places such as Sydney, Melbourne, Brisbane and Perth. While natural population increase in metropolitan areas is partly responsible for this pressure, migration is the key factor. In the period 2011–16, 90 per cent of Australia's new international migrants settled in one of the capital cities, and only 10 per cent in a regional area. Capital cities are also a destination for intra- and interstate migrants. This indicates that the challenges created by demographic change in Australia are spatially differentiated. The implications for people living in metropolitan areas are different from those for people living in regional and rural areas of Australia.

This subtopic will examine some of the social, economic, infrastructure and environmental challenges of population change for a variety of people and places, especially in Queensland, including some of the places referred to in the previous section. For example, Maleny illustrates the effects of demographic change on a rural town experiencing tree-change growth, while Springfield, a newly-developed suburb on the outskirts of a metropolitan area, has experienced rapid growth in the number of families with children.

3.7.1 Metropolitan areas: West End, Springfield

While population growth has city-wide implications, the greatest pressure is often felt in inner and outer suburbs, where growth is usually the most rapid. Inner suburbs have experienced increasing densification through the construction of high-density apartment blocks, while outer suburbs often house young families who are car-dependent and may be involved in long commuting times to schools and places of employment. Table 3.18 shows the growth in population for Australia's major cities between 2006 and 2016. The main implications of this population growth include:

- access to affordable housing
- availability of suitable employment
- availability of health, education and other services
- maintaining accessibility through the development of transport infrastructure, including public transport systems and road networks
- managing growth to maintain urban livability
- pressure on rural and natural environments at urban fringes
- environmental impacts of providing energy, food and water resources
- ensuring social cohesion.

TABLE 3.18 Metropolitan population change, 2006–16

Metropolitan area	2006	2011	2016
Sydney	4 284 379	4 605 992	5 029 768
Melbourne	3 744 373	4 169 103	4 725 316
Brisbane	1 820 400	2 146 577	2 360 241
Perth	1 519 510	1 832 114	2 022 044
Adelaide	1 146 119	1 262 940	1 324 279

Source: Australian Bureau of Statistics

Demographic challenges for inner suburbs

Many inner suburbs have experienced population growth over the past fifteen to twenty years, a consequence of re-urbanisation and the in-migration of younger adults, who are often employed in information and communication technologies and a variety of other professions. Table 3.19 and figure 3.52 provide a selection of demographic statistics for the Brisbane suburb of West End. These statistics illustrate some common demographic features of many inner-city suburbs. One key feature of inner-suburban development and population growth illustrated by West End has been the growth in the number of unit and apartment blocks, and the consequent increase in population density.

TABLE 3.19 Demographic statistics for West End, Queensland

Demographic characteristic	2001	2006	2011	2016	Australia 2016
Population	5680	6206	8061	9474	23 401 886
Population 0–14 years (%)	11.6	11.6	11.5	12.9	18.7
Population 65 years and over (%)	12.5	11.1	8.8	8.8	15.8

(Continued)

TABLE 3.19 Demographic statistics for West End, Queensland *(Continued)*

Demographic characteristic	2001	2006	2011	2016	Australia 2016
Dwelling structure					
Separate house	46.1	43.9	36.9	30.6	72.9
Flat, unit or apartment	47.2	51.3	54.5	63.3	13.1
Semi-detached, row or terrace house, townhouse	5.4	4.1	7.8	5.4	12.7
Occupation	Professionals 35.2%	Professionals 38.1%	Professionals 40.6%	Professionals 42.4%	Professionals 22.2%
	Clerical and administrative workers 26.7%	Clerical and administrative workers 13.3%	Managers 14.2%	Managers 13.0%	Clerical and administrative workers 13.6%
	Managers 7.5%	Managers 10.7%	Clerical and administrative workers 11.7%	Clerical and administrative workers 12.1%	Technicians and trade workers 13.5%

Source: Australian Bureau of Statistics

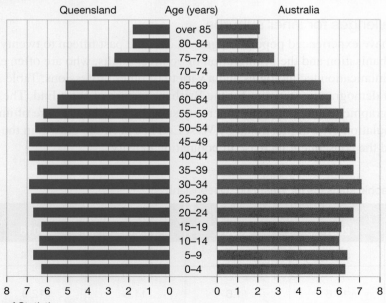

FIGURE 3.52 Population pyramid of West End, 2016

Source: Australian Bureau of Statistics

West End has recently seen a growth in the number of children aged 0 to 14 years. This has put pressure on the resources of the local West End State Primary School. Initially, classroom overcrowding at the school was reduced through the construction of temporary classrooms. Beginning in early 2019, a new campus is being built on land directly opposite the current primary school to cater for the increasing enrolments. The site of this new campus was previously used for light industry. Such changes in land use, known as **brown-field developments**, are common in inner suburbs, and often involve the construction of high-rise apartment blocks

TABLE 3.20 West End State Primary School student numbers

Year	Number of students
2012	716
2013	748
2014	825
2015*	783
2016	854
2017	960
2018	1055
2019	1118

Note: Prior to 2015, state primary school enrollment included Year 7. From 2015, all Year 7 students attended high school.
Source: © The State of Queensland Department of Education 2018

FIGURE 3.53 Construction site for the new West End State Primary School campus

Source: Dr Philip O'Brien

The example of West End State Primary School illustrates the types of challenges that many of Australia's inner-city suburbs face. Families with young children are increasingly finding apartment living close to the city centre appealing, so pressure increases on educational and other social infrastructure, including for health and recreation.

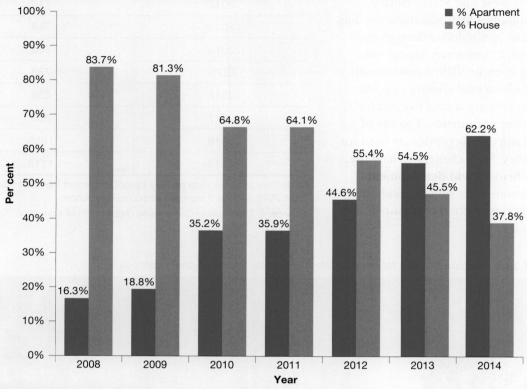

FIGURE 3.54 Change in dwelling type for students enrolled in West End State Primary School, 2008–14

Source: © The State of Queensland 2018

FIGURE 3.55 West End State Primary School and nearby apartment complexes

FIGURE 3.56 Older residences and new apartment buildings adjacent to West End State Primary School

Source: Dr Philip O'Brien

There are other social, economic, infrastructural and environmental consequences of inner suburban population growth and densification. Pressure increases on transport infrastructure, so road congestion worsens, and public transport networks become overcrowded. Often, there is a lack of green space for recreation and the social cohesion of the local community may be put under pressure as in-migration changes the composition of the community. *Jacaranda Senior Geography for Queensland 1* discussed some of the tensions that have occurred in West End as a result of development in the suburb (see section 3.9.3).

FIGURE 3.57 Construction of apartments in inner suburbs increases population pressure on services and infrastructure

FIGURE 3.58 Population growth commonly leads to inner suburban traffic congestion.

Demographic challenges of suburban development on metropolitan fringes

Rapid population growth often occurs on the fringes of Australia's metropolitan areas, frequently leading to urban sprawl. Other than the availability of residential housing, the main factor driving population growth in outer suburbs is housing affordability, although the desirability of larger residential blocks, lower density living and green space is also a consideration for many people with young families. Frequently, the biggest challenge for people living in outer suburbs is the lack of employment opportunities in the local area; long commute times to work are also common. The suburbs might also have infrequent or no public transport, so people become car-dependent.

The suburb of Springfield, which is part of a Greater Springfield development, provides an opportunity to investigate the ways in which the impact of population growth on the outskirts of the Brisbane metropolitan area is being managed. Table 3.21 shows that a high proportion of Springfield's dwellings are separate houses, a typical characteristic of all outer suburbs. Springfield also has a relatively large percentage of its population aged 0–14 years (27 per cent in 2016) compared with 18.7 per cent for Australia as a whole. As in West End, there has been a rapid growth in student enrolments in both Springfield's state primary and secondary schools (table 3.22). However, unlike in West End, this growth in student numbers was anticipated and properly planned for. Currently, there are eleven public and private schools in Greater Springfield to cater for ongoing growth in the number of children.

TABLE 3.21 Demographic statistics for Springfield

Demographic characteristic	2006	2011	2016
Population	5732	6617	6722
Male population (%)	49.2	48.7	49.6
Female population (%)	50.8	51.3	50.4
Population 0–14 years (%)	32.3	29.4	27.0
Population 65 years and over (%)	2.4	3.6	5.4
Median age	26	28	30
Dwelling structure			
Separate house	91.4	92.9	92.5
Flat, unit or apartment	1.8	0	0.4
Semi-detached, row or terrace house, townhouse	6.2	7.1	7.0
Occupation	Clerical and administrative workers 18.9%	Clerical and administrative workers 16.8%	Clerical and administrative workers 16.0%
	Technicians and trades workers 17.4%	Technicians and trades workers 15.8%	Technicians and trades workers 15.0%
	Labourers 11.9%	Professionals 14.3%	Professionals 14.8%
	Professionals 11.9%		

Source: Australian Bureau of Statistics

TABLE 3.22 Springfield State School student numbers

Year	Springfield Central State School	Springfield Central State High School
2014	428	928
2015	548*	1104
2016	664	1192
2017	726	1385
2018	846	1530

Note: Prior to 2015, state primary school enrollment included Year 7. From 2015, all Year 7 students attended high school.
Source: © The State of Queensland Department of Education 2019

Springfield is one of six suburbs in Greater Springfield, Australia's largest master-planned urban development and the first privately built city. Planning has meant that Greater Springfield's growth has occurred in phases, so infrastructure has been built prior to the sale of residential properties. The population of Greater Springfield has gradually grown, from 20 310 in 2011 to 34 000 in 2016, with an anticipated population of 115 000 in 2034.

Unlike many outer suburban developments, Greater Springfield is attempting to create employment opportunities within the community, in the large shopping centre and through the development of three hubs based on health, education and information technology. However, many people still currently commute to work, either in Brisbane or Ipswich. While many use automobiles, there are two rail stations in Greater Springfield, although demand for public transport has exceeded expectations, so parking near the stations has become an issue.

As with all city-fringe **greenfield developments**, there are environmental consequences of the growth of Greater Springfield. Native bushland has to be cleared prior to construction, and drainage patterns may be altered. However, the development does intend to include 30 per cent green space as compensation, and to maintain amenity for the Greater Springfield community.

FIGURE 3.59 Ongoing growth is occurring in Greater Springfield

Source: © Greater Springfield

3.7.2 Regional and rural towns: Maleny, Longreach

Regional and rural towns across Queensland and Australia face contrasting demographic challenges. Many inland towns are confronted with issues of depopulation, demographic decline, ageing, and threats to community social and economic viability. In contrast, coastal and near-metropolitan towns might have to face issues related to strong population growth, such as pressure on infrastructure and the natural environment, and changes to the social and cultural character of the community. In addition, the resource boom in recent years has meant that a number of regional and rural towns have experienced significant population growth followed by demographic decline, a cycle of 'boom and bust'.

Demographic challenges for tree- and sea-change towns

A tree change is a move to a vegetated rural location from a major urban centre. Similarly, a sea change is a move from a major urban centre to a rural location near the coast. Both moves are usually an attempt to escape the stress associated with urban lifestyles. Maleny has experienced rapid population growth over the past fifteen to twenty years, much of which has been driven by tree changers pulled by the town's relative proximity to Brisbane and the Sunshine Coast, and its favourable climate and rural amenities. As the increasing percentage of the population aged 65 years and older suggests, a large number, although not all, of Maleny's immigrants have been looking for an attractive place to retire. This inflow of people has brought development pressure for Maleny, and has also resulted in a number of protests against proposed developments.

TABLE 3.23 Demographic statistics for Maleny

Demographic characteristic	2001	2006	2011	2016
Population	1,104	1,294	2,489	3,743
Male population (%)	45.7	45.2	44.8	47
Female population (%)	54.3	54.8	55.2	53
Population 0–14 years (%)	22.8	21.5	17.0	13.4
Population 65 years and over (%)	19.3	20.0	26.9	36.3
Median age	N/A	42	51	57

Source: Australian Bureau of Statistics

As Maleny's population has grown, its urban footprint has expanded, with residential development occurring on the fringes of the town. From the 1980s, some of this residential growth included large houses built on half- to one-hectare blocks. Given Maleny's age profile, it seems likely that there will be an increasing demand for aged-care services and for retirement infrastructure.

FIGURE 3.60 Maleny's ageing population will need more retirement homes, such as Maleny Grove Life Village

Source: Christopher Frederick Jones

Proposed developments in the urban centre of Maleny have caused controversy. A proposal in 2005 for the establishment of a Woolworths supermarket was opposed at the time for environmental, economic and social reasons by over 70 per cent of the local population. Residents were concerned that the development site was a significant platypus habitat, that local businesses would be affected and that the village character of the town would be altered. Despite the protests, the Woolworths development proceeded, and the supermarket opened in April 2006.

Protests against the construction of a petrol station and convenience store near the Maleny Primary School were more successful and the development proposal was abandoned. Also abandoned was a proposal for a 'TT Style' motorcycle race for the region, which was opposed by Maleny residents because of noise, safety and access concerns, and the belief that the race did not fit with existing tourist activities.

Demographic challenges for inland towns

The demographic challenges faced by the central-west Queensland town of Longreach contrast with those faced by Maleny. Unlike Maleny, but like many of Australia's other inland towns, Longreach is currently experiencing population decline, as the figures in table 3.24 show. Longreach's population decline is partly explained by the 2012–19 drought, which had a significant impact on the region's pastoral industry. However, even without the serious challenges posed by drought, many rural towns have experienced long-term demographic decline, as younger people leave the land, and the agricultural workforce ages.

FIGURE 3.61 The town of Longreach

TABLE 3.24 Demographic statistics for Longreach

Demographic characteristic	2001	2006	2011	2016	Queensland 2016
Population	3648	2976	3137	2738	4 703 193
Male population (%)	50.3	49.0	47.9	47.8	49.4
Female population (%)	49.7	51.0	52.1	52.2	50.6
Population 0–14 years (%)	21.5	23.0	25.1	19.2	19.4
Population 15–24 years (%)	12.6	14.9	12.1	14.0	13.0
Population 65 years and over (%)	15.6	13.5	12.0	16.1	15.2
Median age	N/A	33	32	38	37

Source: Australian Bureau of Statistics

One consequence of Longreach's declining population is a fall in enrolments in the Longreach State School and Longreach State High School (table 3.25). Demographic decline also resulted in the closure of the Longreach Student Hostel in 2015, when it had only five students, down from 18 in 2009. By 2015, the hostel, which had operated for thirty-six years as a residential facility for high school students from surrounding rural areas, had become economically unviable.

TABLE 3.25 Longreach state schools student numbers

Year	Longreach State School	Longreach State High School
2014	260	201
2015	229*	225
2016	201	221
2017	198	187
2018	191	177

Note: Prior to 2015, state primary school enrollment included Year 7. From 2015, all Year 7 students attended high school.
Source: © The State of Queensland Department of Education 2019

The Longreach Pastoral College is another of Longreach's educational facilities forced to close because of falling student demand and financial losses. The college opened in 1969 to train students entering the pastoral industry, but had only nine graduating students when it closed in 2019. A report to the Queensland government about the operation of the Longreach Pastoral College and the Emerald Agricultural College did point out that, as new industries and new jobs emerge across the rural and regional economy, the future for facilities such as these could involve education and training courses in areas such as tourism and hospitality, health and

FIGURE 3.62 Longreach Pastoral College

Source: Richard Durham / AAP

community care, IT literacy, solar farming and drone/robotics/GPS technologies for agriculture, pastoralism and mining.

Another consequence of Longreach's declining population has been the decline in real estate prices and an increase in the numbers of residential properties for sale. In Longreach, Mt Isa and Birdsville, property prices declined by 14 per cent in 2018, and by 8 per cent in the last three months of 2018 alone. In 2016, Longreach had 200 properties for sale and a large number of vacant rental properties. Given the relative importance of primary and secondary education for employment in Longreach (table 3.26), declining school numbers may also have an impact on the retention of teachers and their families in the town, so housing demand may continue to fall and overall population continue to decline.

TABLE 3.26 Top five industries of employment, Longreach, 2016

Industry of employment	Longreach	Australia*
Hospitals (except psychiatric hospitals)	7.1	3.9
Local government administration	5.2	1.3
Primary education	4.0	2.2
Supermarkets and grocery stores	3.9	2.4
Secondary education	3.5	1.7

Note: Australia's top five industries of employment in 2016 did not include local government administration or secondary education. The top 5 were: Hospitals (except psychiatric hospitals), supermarket and grocery stores, cafes and restaurants (2.4 per cent), primary education, and aged-care residential services (2.0 per cent).

Source: Australian Bureau of Statistics

FIGURE 3.63 Significant locations in central-west Queensland

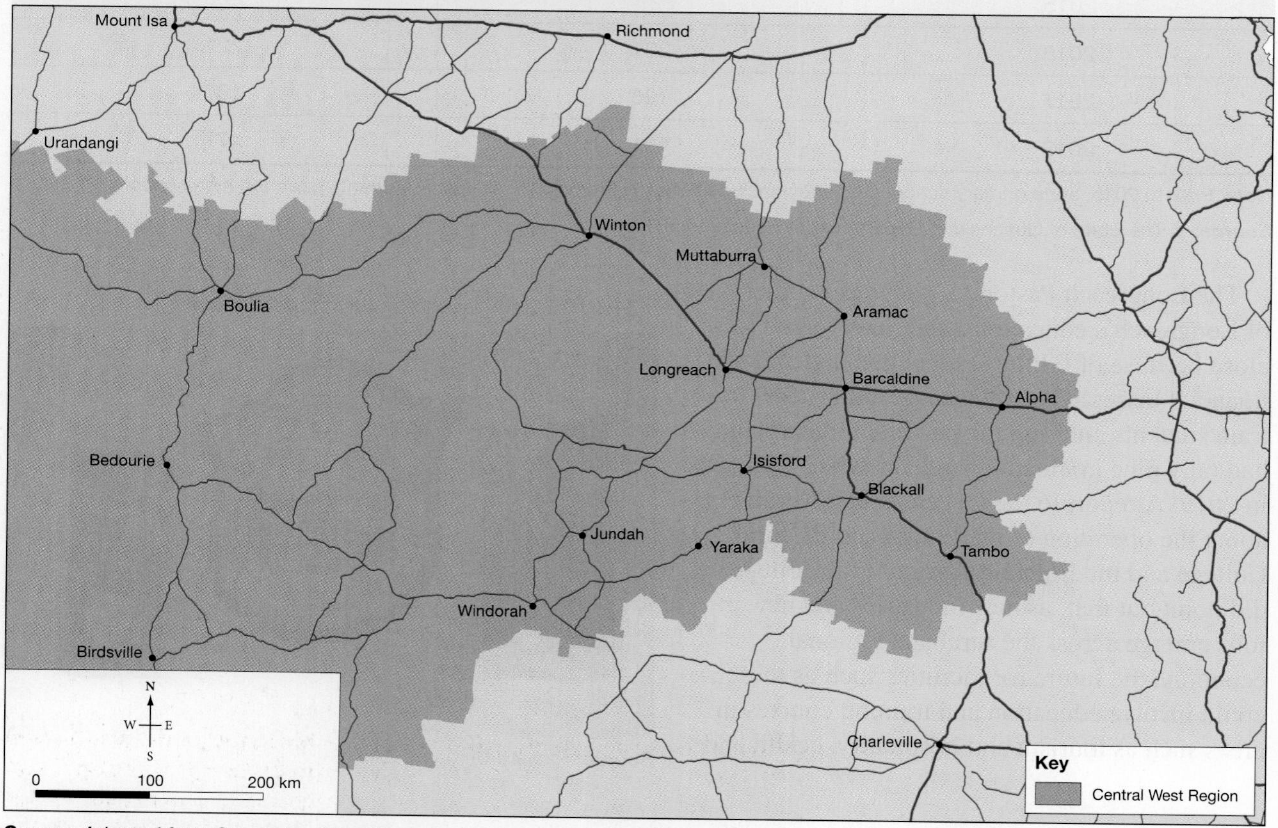

Source: Adapted from Queensland Health

Such declines in property prices, unsold houses and vacancies are an indication of the economic pressures faced by Longreach. Drought and the lack of jobs have resulted in out-migration of people. This has reduced demand for goods and services in the town, leading to reduced demand for labour and the potential closure of some businesses. Consequently, there is further population decline and further economic decline, creating a downward cycle.

Activity 3.7: Identifying and responding to population challenges at a local scale

Describe and analyse data

This subtopic has outlined some of the demographic features of and the challenges faced by several places in Queensland. This activity provides you with an opportunity to practise some of the skills needed for your geographic inquiry and data report assessment task.

Choose one of the four case studies presented in this subtopic to complete the following tasks.
1. Describe the demographic characteristics of the selected place. You should refer directly to the demographic data provided in the tables. You might also like to gather additional data from the Australian Bureau of Statistics and other sources.
2. Analyse the demographic data and describe the demographic challenges faced by the place.
    As part of your analysis, you should include some mathematical calculations (e.g. of the rate of population growth or decline and the dependency ratio). You should use spatial and information and communication technologies as part of your analysis of the data.
    You could construct line or bar graphs, or find data to use in constructing a catchment area map for one of the schools mentioned in the text. Google Maps or the ABS online mapping tool **TableBuilder** can be used for constructing maps.

Explain and discuss demographic processes

3. Explain the geographic and demographic processes that have contributed to the demographic challenges you have identified.
4. Discuss the current and likely future impacts of the demographic challenges faced by your selected place.
5. Suggest some possible actions that your selected place could make in response to the demographic challenges it faces. To what extent will your proposed actions be economically, socially and environmentally sustainable?

For further assistance a Practice Data Report task and a scaffolded Data report are provided in the Resources tab of your eBook.

Resources

Digital documents Practice Data Report task (doc-31438)
Scaffolded Data Report booklet (doc-31439)

Video eLessons SkillBuilder: Drawing a line graph (eles-1635)
SkillBuilder: Creating a simple column graph (eles-1639)

Interactivities SkillBuilder: Drawing a line graph (int-3131)
SkillBuilder: Creating a simple column graph (int-3135)

Weblink ABS TableBuilder tool

3.8 Review

3.8.1 Chapter summary

This chapter has covered some key points about Australia's population patterns and trends, including what is changing about our population and why.

Demographic concepts
- Demography involves statistical analysis of characteristics of populations such as size, composition and distribution across space, and the processes through which populations change over time.
- Key demographic concepts include birth rate, death rate, fertility rate, life expectancy, age–sex structure, migration rate, total population growth and rate of natural increase.
- There are significant differences in countries and regions around the world in birth, death and fertility rates.
- Population pyramids show detailed data on the age of the population of countries and regions, as well as the male–female breakdown of the population.
- The age structure of a population illustrated by population pyramids often has implications for planning by governments, non-government organisations, businesses and individuals.

Changes in population
- Changes over time in birth rates and death rates in a country or region have an impact on the rate at which its population changes.
- The rate of natural increase or decrease in a country's or region's population depends on the relationship between birth and death rates.
- If birth rates are much higher than death rates for a lengthy period, the population will grow rapidly; if the death rate exceeds the birth rate, the population may decline. If the rates are close, the population will change slowly.
- Migration rates also have an impact on the rate at which a country's or region's population changes.
- Australia's population has more than doubled since the 1960s, although its relative contribution to global population growth is relatively insignificant.

Factors affecting population change
- Factors influencing population change in Australia include advances in health care and life expectancy, birth rates and the changing role of women in society, the impact of disease on death rates, migration policies over time, and amenity.

Population patterns and trends
- There is a high degree of geographical concentration in Australia's population with over 70 per cent of the population located in the largest cities.
- Most of Australia's population lives on or near the coast, and a large area of Australia is sparsely populated.
- There are both geographical and historical reasons for the distribution pattern of Australia's population.
- There are differences in the age structure of the populations of capital cities and other major urban areas compared with smaller towns and villages and rural areas.
- There are differences in the age structures of the states and territories, as well as between places within the states and territories.
- There is little difference between the capital cities and the rest of Australia in sex structure.
- New South Wales and Victoria, Australia's most populous states, have the highest proportions of overseas-born people.
- A very high percentage of migrants to Australia lives in the capital cities, particularly Sydney and Melbourne.

- The distribution of migrants across the capital cities varies, both in terms of location and countries of origin.
- In Queensland, Brisbane has the highest number of overseas-born people, followed by the Gold Coast, but all parts of Queensland have some migrants.
- In 2016, the largest number of Aboriginal and Torres Strait Islander people lived in New South Wales, with Queensland second.
- Indigenous Australians make up over a quarter of the population of the Northern Territory, and also outnumber overseas-born people in that Territory.
- In comparison to the rest of Australia's population, Indigenous Australians are relatively young, with a median age of 23, and 34 per cent of the population aged below 15 years.
- Life expectancy and death rates for Indigenous Australians are worse than for non-Indigenous Australians

Local area population patterns
- The demographic characteristics of Australia's states and of regions within states can vary, as can the demographic character of smaller local areas such as regional cities, country towns and the suburbs of cities.
- Demographic challenges can occur as populations change spatially and over time.

Implications for people and places of demographic change
- Demographic challenges can include pressure from population growth on services, transport and other infrastructure, housing and employment, as well as pressures from ageing and population decline.
- Inner city areas might face challenges of densification and lack of schools and other services, while outer suburbs may face challenges of limited employment opportunities and long commutes to school and work.
- Regional and rural towns across Queensland and Australia face contrasting demographic challenges.
- Inland towns might face challenges of depopulation, demographic decline, ageing, and threats to community social and economic viability.
- Coastal and near-metropolitan towns might face challenges related to strong population growth, such as pressure on infrastructure, services and the natural environment, and changes to the social and cultural character of the community.

3.8.2 Key questions revisited

You should now be able to answer the following questions.
- What is demography? What are the key demographic concepts?
- What are Australia's population patterns and trends?
- What demographic processes are responsible for changes in population in Australia?
- What factors influence changes in population in Australia, both spatially and over time?
- What are the implications of population change for people and places in Australia?
- What are the demographic characteristics of a particular place in Australia?
- What challenges occur because of demographic change in a particular place in Australia?
- What actions might be taken to manage these challenges?

4 Global population change

4.1 Overview

4.1.1 Introduction

In this topic you will examine patterns and trends in world population change, the factors that have contributed to spatial variations in population growth and decline, and the relationships and implications for people and places as a result of this. You will also investigate the changing distribution of the world's population, the processes that have led to movements of people across the world and the impact this has had on places of origin and destination.

FIGURE 4.1 The size of a country's population can determine the decisions the government makes.

4.1.2 Key questions

- Why is accurate and reliable census data useful to governments?
- What trends and patterns can be identified in global population change?
- Which demographic concepts are critical in explaining population change?
- How can models help us understand the past and possible future changes in population across the world?

- What spatial patterns of population distribution and density result from population change?
- How do countries manage the challenges of changing characteristics of their populations such as ageing?
- What are the positive and negative impacts of migration of people on places of origin and destination?

4.2 Global patterns of population growth

4.2.1 The population census

A national population census is a count of the number of people living in a country at a particular point in time. The earliest known population census was conducted in ancient Babylonia (now part of modern-day Iraq) in 3800 BC but the oldest existing census was conducted in 2 AD in China during the Han Dynasty. At the time, the Chinese population was estimated to be almost 60 million. Today, governments throughout the world systematically conduct a census to determine not only the total number of people living in their country but also to find out the characteristics of the population, such as age and gender. This data enables governments to make relatively accurate future population forecasts and enable them to plan for population growth or, in some cases, population decline.

The United Nations (UN) uses census data from national governments to make global and regional forecasts of population change in terms of numbers and characteristics. UN forecasts have been a useful tool for national and international organisations for predicting the impact of population change on resource use and **sustainability**. Errors in UN forecasting have been surprisingly small in countries with large populations.

The accuracy of some national census data has not been without controversy. The newly independent Nigerian government held its first census held in 1962 and preliminary results showed that the north part of the country had lost its majority share of the total population. This was disputed by its leaders so a second census held in the following year, during which an additional 8.5 million people were discovered in the north. A third census held in 1973 was so hotly disputed that the government simply declared the result null and void. These early post-colonial Nigerian censuses were important because **population size** determined parliamentary representation and revenue allocation.

4.2.2 Global population change

The global population is currently over 7.7 billion, but this growth has experienced several distinct phases over the last 300 years, as shown in figure 4.2. Until the late-1700s, the global population grew very slowly — the annual growth rate was less than 0.1 per cent. By the early 19th century the global population had passed the billion people mark and global population growth accelerated with an average annual rate of 0.5 per cent, which led to a dramatic increase in population during the first half of the 20th century. It had taken 130 years for the global population to double from the mid-18th century but within 50 years it had doubled again. Global population growth peaked in the early 1960s at an annual growth rate of just over 2 per cent but since then it has fallen. Despite the slackening of the demographic growth rate, the world's population has increased by one billion every 12 years since 1962. Consequently, the UN has estimated that the global population is likely to grow to nearly 10 billion by 2050.

Global patterns conceal differences in growth rates between countries and regions, particularly in terms of their different levels of economic development (see figure 4.3). Countries in the **developed world** were the first to experience significant improvements in **life expectancy**, which in turn brought about a steady decline in **death rates** while **birth rates** remained high. This resulted in much higher population growth rates than in any other part of the world. The population growth rate of so-called **newly industrialised countries (NICs)** generally peaked in peaked in the mid-20th century whereas the **least developed countries** experienced this peak towards the end of the 20th century. In recent decades, population growth rates have generally been in decline in countries at all stages of development.

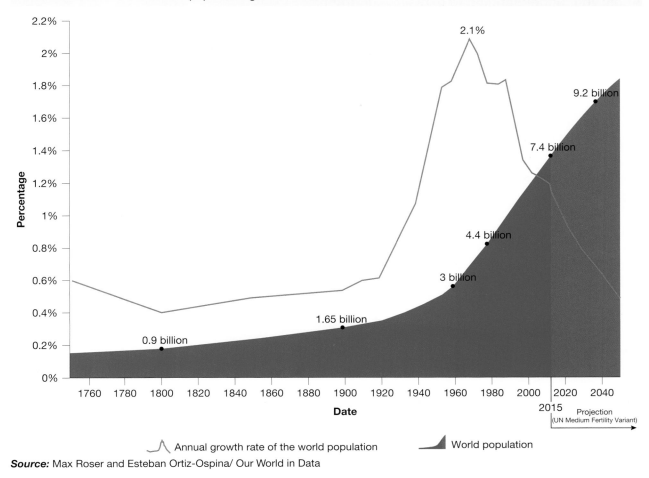

FIGURE 4.2 World and annual population growth 1750–2050

Source: Max Roser and Esteban Ortiz-Ospina/ Our World in Data

Activity 4.2a: Interpreting world population data

Explain and analyse population data

1. Study figure 4.3.
   (a) List the world regions in rank order according to their share of global population in 1820, 1900 and 2015.
   (b) How has the rank order for the six world regions changed since 1820?
   (c) Compare the growth of Asia's population since 1820 with one of the other world regions. How is Asia's pattern of demographic growth similar and different?
   (d) Access **Gap Minder** by going to the Resources panel. Click on the Ranks icon, which shows countries of the different regions in rank order. Select the required year and use the information to explain Asia's pattern of demographic growth.

Synthesise the information and communicate using graphs

2. (a) Using data from table 4.1, construct either three six column bar graphs or three pie charts to illustrate the changing distribution of the world's population since 1950.
   (b) Describe the pattern of change between 1950 and 2050.

TABLE 4.1 Changing distribution of the world's population since 1950

World regions	Percentage share of global population		
	1950	2019	2050
Africa	9.0	17.1	25.9
Asia	55.4	59.4	53.8
Europe	21.7	9.6	7.3
Latin America and the Caribbean	6.7	8.5	8.0
North America	6.8	4.8	4.4
Oceania	0.5	0.5	0.6

Source: © Copyright Worldometers.info; United Nations, Department of Economic and Social Affairs, Population Division (2017). World Population Prospects: The 2017 Revision, custom data acquired via website

Apply your understanding to explain your graphs

3. Explain the predicted distribution of the world's population in 2050 shown in your previous answer, given the assumption that the trends shown in figure 4.4 continue.

FIGURE 4.3 Population growth by world regions

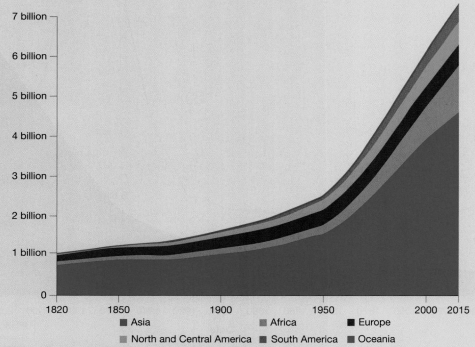

■ Asia ■ Africa ■ Europe
■ North and Central America ■ South America ■ Oceania

Source: Max Roser and Esteban Ortiz-Ospina/Our World in Data

FIGURE 4.4 Population growth rate, 2015

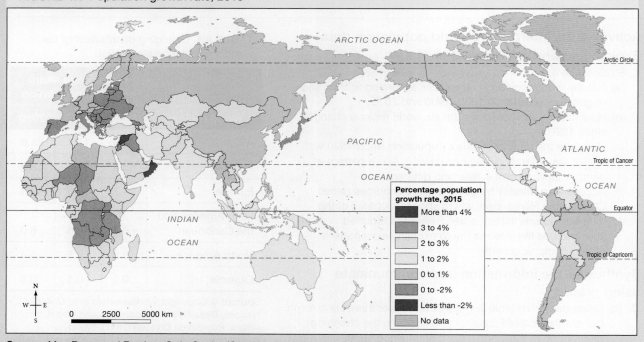

Source: Max Roser and Esteban Ortiz-Ospina/Our World In Data

Resources

Video eLessons SkillBuilder: Creating a simple column graph (eles-1639)

SkillBuilder Constructing a pie graph (eles-1632)

Interactivities SkillBuilder: Creating a simple column graph (int-3135)

SkillBuilder: Constructing a pie graph (int-3128)

Weblink Gap Minder

4.2.3 The future world population size and composition

The UN has produced projections for global population growth rates since the 1950s. Using current data, the UN makes assumptions about how the population may change to produce what is the most likely outcome. Traditionally, projection variants have been produced that show a range of possible population growth forecasts. The medium variant is the most likely population growth forecast whereas the high and low variants are less certain and have upper and lower limits of reasonable projections (see figure 4.5).

There are three factors that determine the future global population size and composition. These are changes in:
- fertility levels
- **mortality rates**
- **age–sex structure**.

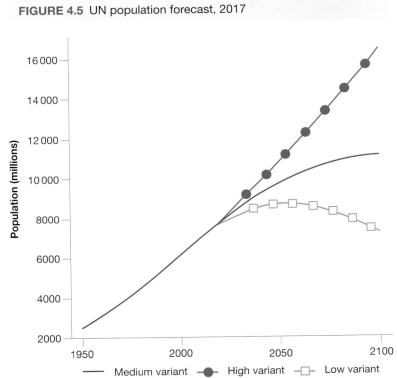

FIGURE 4.5 UN population forecast, 2017

Source: World Population Prospects 2017 by United Nations / Population Division. Copyright © 2017 United Nations. Used with permission of the United Nations.

Fertility rate

The **world fertility rate** is the number of children born per woman of childbearing age. This rate has declined from more than 5.0 in the mid-1960s to just under 2.5 currently. If the rate falls below 2.1 it will contribute to population decline. To maintain what is known as **replacement fertility**, which is the rate required for a population to replace itself from one generation to the next (approximately 2.1 births per woman), there must be one surviving daughter at each pregnancy. This is because there is a slight gender imbalance at birth, skewed towards boys, and the chances of survival from birth to the reproductive ages are less than 100 per cent.

Fertility rates vary between world regions and between countries (see figure 4.6). Least developed countries will likely continue to have relatively high levels of fertility but they are declining. This is in sharp contrast to the increasing number of countries, which account for nearly half the world's population, that have fertility rates below replacement level. Unlike other regions in the developed world, the fertility rate in Europe it is expected to rise from 1.6 to 1.8 children born per woman of childbearing age by 2050.

FIGURE 4.6 Total fertility, medium projection, 2020–25

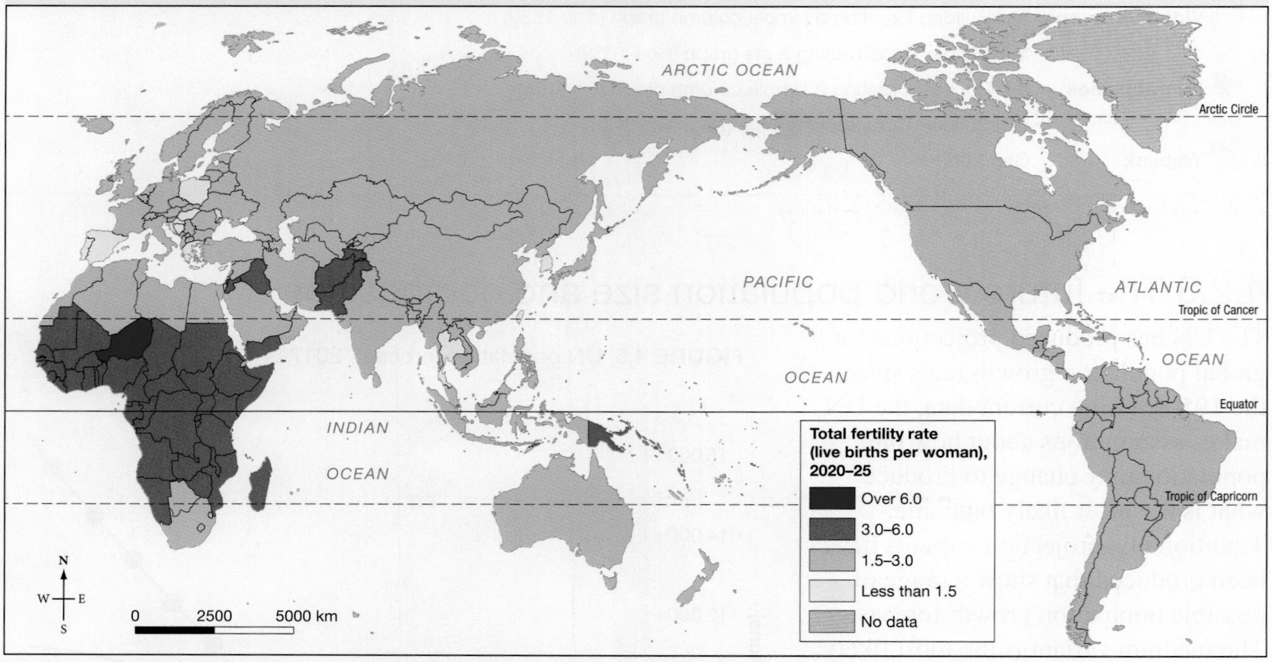

Total fertility rate (live births per woman), 2020–25
- Over 6.0
- 3.0–6.0
- 1.5–3.0
- Less than 1.5
- No data

Source: From World Population Prospects: The 2017 Revision, by United Nations–Department of Economic and Social Affairs, © 2017 United Nations. Used with the permission of the United Nations.

UGANDA'S POPULATION GROWTH IN AN AGEING WORLD

Although the average African fertility rate fell from 5.1 in 2005 to 4.7 in 2015, the most youthful and fastest growing populations are to be found in this region. In the east of Africa, Uganda is one of the fastest growing countries in the African region. Although Uganda's fertility rate has fallen from a high of 7.1 in 1994, its population is growing at a rate of more than 3 per cent per annum. Therefore, its population is likely to mushroom from 28 million in 2019 to 130 million by 2050.

Vision Uganda 2040 was launched in 2007 to set out a development path that would create a modern and prosperous country within 30 years. Uganda's growing population was seen as an asset, sometimes known as a demographic dividend, and not a liability holding the country back. This view was based on the experience of Asian countries that had achieved remarkable economic growth despite possessing large populations, and partly explains what some observers believe is the Ugandan government's lack of commitment to family planning.

FIGURE 4.7 Systems diagram* of population dynamics in Uganda

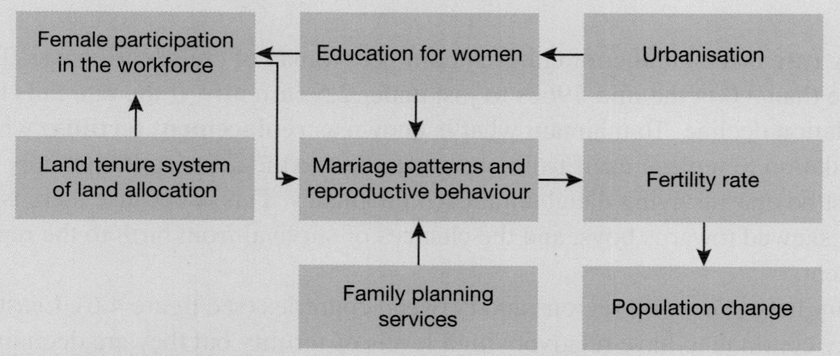

Note: A systems diagram is a flow diagram comprising inputs, outputs, stores and flows. They help us understand how complex systems work and how the different components of a system are inter-related, particularly the way a change of one factor impacts on the rest of the system.

The Demographic and Health Report for Uganda, published in 2011, revealed an insight into the rate of pregnancy and the use of contraceptives, and the differences in family planning between rural and urban areas. The report showed high levels of pregnancy among teenage women — almost one fifth of all Ugandan women between the ages of 15 and 19 have already had one child. It also showed that the use of contraception is particularly low, with only a third of married women and half of single unmarried women using some method of contraception. A barrier to greater use of contraceptives is the common and misinformed belief among men that they can cause infertility and cancer. The report noted differences between rural and urban areas, with access to family planning services and education for women much greater in towns and cities than in rural areas. This is significant because the median age of first birth for women is two years higher for those who have completed secondary education.

The Ugandan government is introducing family planning initiatives, such as Family Planning 2020 Partnership, which has received support from the United Nations Population Fund (UNFPA). The government has committed to providing universal access to family planning and reduce the unmet need for family planning to 10 per cent by 2022. In 2018, the UK government signed a five-year agreement with the UNFPA to provide additional funding for Uganda, its former colony.

In rural Uganda, 80 per cent of the land falls under the customary land tenure system. The Lands Ministry is responsible for granting certificates of customary ownership, which allows it to regulate family use of communal lands to control degradation. Because customary ownership is dominated by large families, the ministry is now encouraging the allocation of land to girls so they can earn an income from the family land. This is seen as an attractive alternative to early marriage, which can be prompted by the expectation of acquiring land from the husband's family.

Activity 4.2b: Interpreting population pyramids and systems diagrams

Explain the population pyramid

1. Explain why the population pyramid for Uganda (figure 4.8) is indicative of a fast-growing population.

Analyse the data and apply your knowledge of fertility rates

2. With the aid of figure 4.7, what is the likely impact of increased female participation in the workforce and increased urbanisation on Uganda's population growth? Explain your answers.

FIGURE 4.8 Uganda population pyramid, 2017

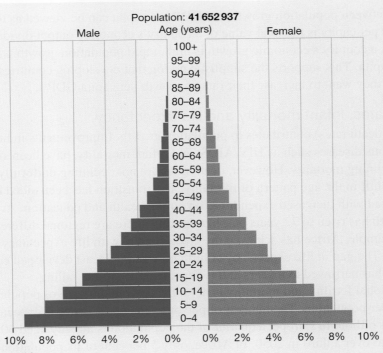

Source: PopulationPyramid.net

Population growth and levels of economic development

Total fertility is a key indicator of population growth. A comparison of figures 4.6 and 4.9 suggests that high population growth rates will continue in developing regions and account for most of the world's population increase to 2030. This is in sharp contrast to developed countries, which will experience sluggish growth of less than 5 per cent.

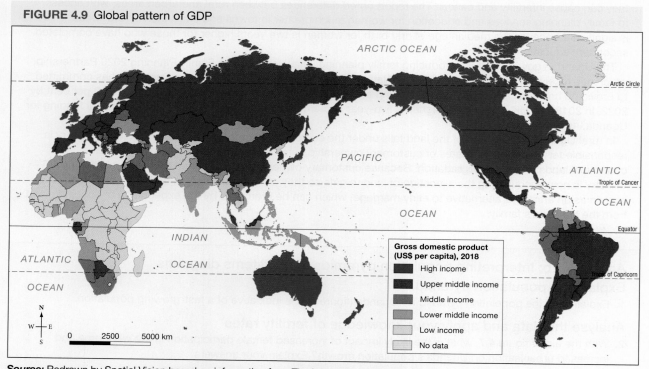

FIGURE 4.9 Global pattern of GDP

Source: Redrawn by Spatial Vision based on information from The International Monetary Fund

The relationship between population growth and GDP per capita can be viewed as too simplistic. This is because GDP per capita is the total value of a country's economic output divided by population. Consequently, unless a country's economic growth is high, rapid population growth will result in a low value for GDP per capita. This supports the simplistic notion that developing countries must limit their population growth if they want to increase their rate of growth per capita GDP.

Mortality: death rates, infant mortality and life expectancy

Declining mortality (death rates) contributes to population growth. Improvements in health, particularly the decline in infectious diseases such as HIV/AIDS, and infant mortality, have been significant factors in recent years to declining mortality. However, death rates are not declining uniformly across all age groups, leading to a shift in the age pattern of mortality. The transition has been aided by socioeconomic development associated with increased expenditure on public health and education. Average global life expectancy at birth has risen to 72 years but, like fertility, there are regional differences in rates of improvement. For example, Africa has achieved the greatest gain, with life expectancy rising by more than 6.5 years since 2000. The gap in life expectancy at birth between the least developed countries and other developing countries has narrowed to 8 years and is expected to continue falling.

The reduction in global fertility and mortality is not only slowing the pace of population growth, it is also producing an older population. The number of persons aged 60 or above is expected to more than double by 2050 to 2.1 billion. In Europe, a quarter of the population is already aged 60 years or over and is predicted increase to over a third in 2050. In Africa, only 5 per cent of its population is aged 60 or over and is expected to rise to only 9 per cent in 2050.

When discussing demographic measures such as fertility and death rates, it is useful to remember these formulas covered in chapter 3.

Crude birth rate:

births per 1000 = $\frac{\text{births per year}}{\text{total population}} \times 1000$

Crude death rate:

deaths per 1000 = $\frac{\text{deaths per year}}{\text{total population}} \times 1000$

Rate of natural increase as a percentage:
(crude birth rate − crude death rate) × 100/1000

Thomas Malthus was an English scholar and economist who developed theories about population growth. In *Essay on the Principle of Population*, first published in 1798, Malthus formulated probably the best known theory on population growth and wellbeing — that rising average incomes would be driven down by population growth, due to the improvements in income and, therefore, wellbeing being offset by diminishing returns of labour in production.

The **Industrial Revolution**, which began in the mid-18th century, marked a major turning point in the relationship between population growth and income per capita. Up until the Industrial Revolution, population growth appeared positively related to the level of income per capita. Technological progress was very slow as was growth in population, which is estimated to have been about 0.3 per cent. With an average economic growth rate of 0.4 per cent, per capita GDP growth averaged 0.1 per cent. The arrival and progress of the Industrial Revolution increased income and population growth, as well as the supplies of food, something Malthus could not have predicted. Subsequently, the positive relationship between income per capita and population growth reversed as population growth slackened and income per capita continued to increase.

FIGURE 4.10 Thomas Malthus

Source: Thomas Robert Malthus. Mezzotint by John Linnell, 1834. Credit: Wellcome Collection. CC BY

Age–sex structure

The age–sex structure of a population defines the relative numbers of young and old as well as the balance of males and females, which in turn influence the overall number of births and deaths. The ratio of males to females born is 105:100 and the replacement level is set at 2.1. Moreover, even if world fertility is at replacement level and mortality remains constant, global population will still increase as a result of **population momentum growth**. This means that the higher the percentage of young people (especially those under age 15), the more the population will continue to rise as this large segment of the population enters its reproductive years (15–49). However, once this group moves beyond childbearing age the momentum will decrease, and the population can begin to stabilise so that births and deaths balance, assuming the fertility rates remain at or below replacement levels.

The 2017 Revision of the World Population Prospects published by the United Nations includes the population momentum growth (the momentum variant). Its value is the ratio of ultimate population size to current population size. The UN has predicted that the global population will continue to increase to nearly 10 billion in 2050 and then stabilise. Figure 4.12 shows how the impact of population momentum diminishes over time. The momentum variant levels out after 2060.

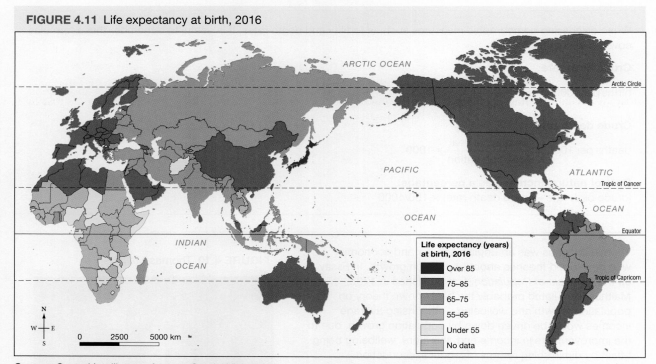

FIGURE 4.11 Life expectancy at birth, 2016

Source: Central Intelligence Agency, Spatial Vision

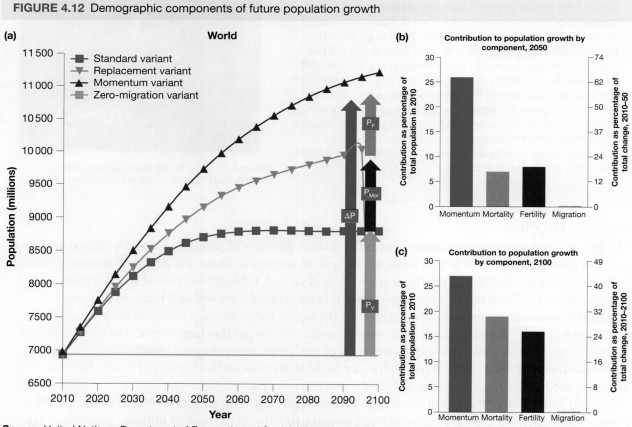

FIGURE 4.12 Demographic components of future population growth

Source: United Nations, Department of Economic and Social Affairs, Population Division (2017). World Population Prospects: The 2017 Revision.

The contribution of the population momentum to population growth will vary between countries. Those with young age structures and total fertility at replacement level are projected to grow because births produced by a large number of females of reproductive age will exceed deaths. For example, Malaysia reached replacement level fertility in 2000. However, with a quarter of its population under age 15 it is expected to grow until 2070 (see figure 4.13). In contrast, for countries with high fertility levels, such as those in Africa, the impact of population momentum in the coming decades will be small because most of the projected growth will be driven by the fertility level rather than the age of the population.

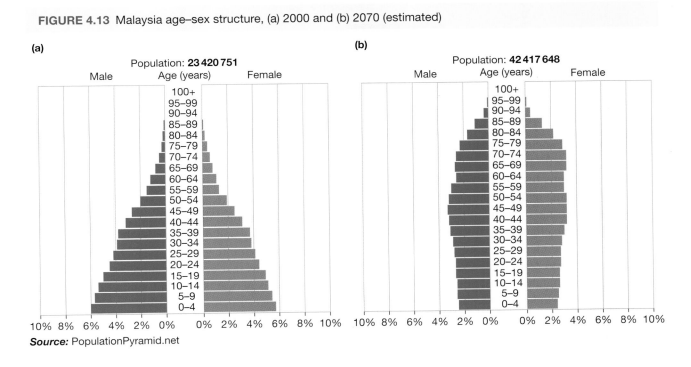

FIGURE 4.13 Malaysia age–sex structure, (a) 2000 and (b) 2070 (estimated)

Source: PopulationPyramid.net

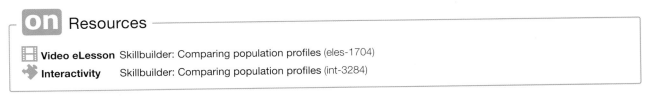

Video eLesson Skillbuilder: Comparing population profiles (eles-1704)
Interactivity Skillbuilder: Comparing population profiles (int-3284)

4.2.4 Theories of population growth and economic development

Rostow's stages of growth

Modernisation theory was developed in the mid-20th century primarily to provide an alternative to communism as a solution to poverty in the developing world. It touches on the link between population growth and economic development, and argued that low income countries needed to follow the same path to development as the West. In other words, they needed to adopt Western cultural values and industrialise.

The most well-known modernisation theory was developed in 1960 by Walt Rostow, an American economist. Rostow's 'Five Stages of Economic Growth' suggested that after initial capital investment, countries would embark on an evolutionary process lasting about 60 years, in which they would move up through five stages of development (see figure 4.14).

Rostow believed that every country would lie somewhere on his development spectrum and during the process of economic development would progress through each stage. These stages had specific characteristics.

FIGURE 4.14 Rostow's Model of Economic Growth

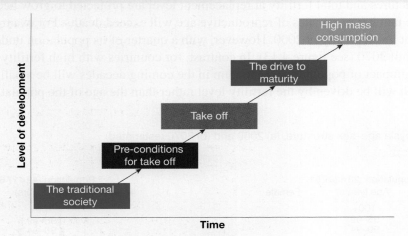

- **Traditional society** is characterised by a labour-intensive agrarian economy and a relatively static population with limited technology.
- **Preconditions for take off** is the stage in which western development aid and investment would assist improvements in agriculture and economic infrastructure, such as roads, and therefore encourage further overseas investment by foreign companies. It could be assumed that this would lead to demographic change, particularly improvements in life expectancy and declining death rates.
- **Take off** is a relatively short industrialisation period of rapid economic growth in which a new urbanised entrepreneurial and middle class of consumers emerges that generates more wealth as a result of cumulative causation. Demographic growth during this period is high and migration from rural areas to the new industrial cities picks up momentum.

Cumulative causation, also known as the **multiplier effect**, is an economic snowballing process by which investment in new industry creates new employment opportunities, which in turn generates more consumers and increased demand for local goods and services. The local economy prospers, which attracts migrant workers from elsewhere.

- **Drive to maturity** takes place over a long period of time. With increasing standards of living and a diversifying economy, social changes such as later marriage bring about a slackening of population growth, a direct consequence of a declining birth rate.
- **Age of high mass consumption** is characterised by a highly urbanised consumer economy. At the time of writing his theory Rostow believed that Western countries, most notably the United States, had reached this last stage. Demographically, the US population growth rate was in decline, having peaked in 1910 (see figure 4.15). In 1960, the US recorded a birth rate of 23 000 and a death rate of 9000 (see figure 4.16).

FIGURE 4.15 US population growth, 1910–2010

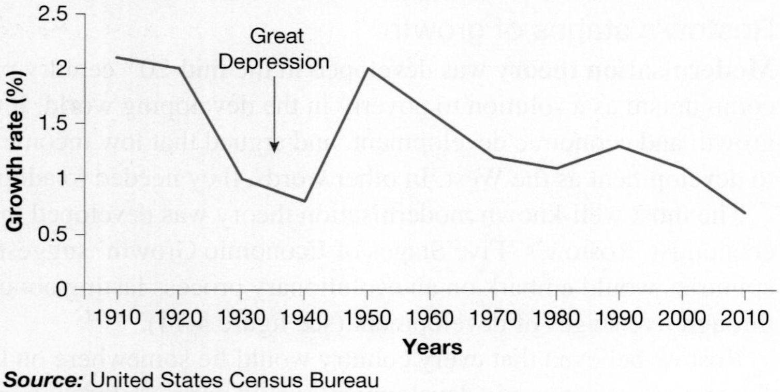

Source: United States Census Bureau

Figures 4.15 and 4.16 show that US population growth rates in the 20th century appear to have been influenced by the health of the economy. The Great Depression, which lasted from 1929 to 1939, lead to unprecedented levels of unemployment, peaking in 1933 at 25 per cent. This had a significant impact on population growth. More recent economic downturns, such as the oil crisis and recession in 2008, have been linked to population growth declines. This has led observers to conclude that whenever the US economy performs well, population growth picks up but when it falters, population growth eases.

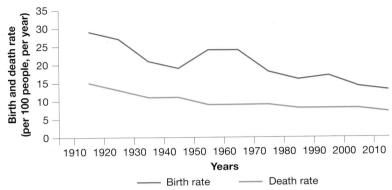

FIGURE 4.16 US birth and death rates, 1910–2010

Source: U.S. National Center for Health Statistics, *Vital Statistics of the United States*, annual; *National Vital Statistics Reports* (NVSR) (formerly *Monthly Vital Statistics Report*).

Wallerstein's world systems theory

In 1974, Immanuel Wallerstein's *The Modern World System* was published. Wallerstein was an American sociologist who produced a theoretical framework comprising four stages in which he attempted to explain how the modern capitalist world economy evolved from the age of feudalism to the present day. Wallerstein argued that through international trade a world economic system had developed with increasing economic and social disparities. He identified three different types of region, each of which possessed certain political, economic and demographic characteristics (see figure 4.17). These regions were:
- **core** technologically-advanced countries such as the US that export capital intensive products to the semi-peripheral and peripheral regions.
- **semi-peripheral** second tier industrialised countries such as Brazil that are not as advanced as those of the core.
- **peripheral** agricultural countries, such as those found in Africa, that export commodities to the core and semi-peripheral regions.

Wallerstein's theory centred on trade between countries. In the early stages of development, the collapse of feudalism in Europe had ushered in a new world economy with empires, which initiated a flow of raw materials from the periphery to the north-west European core region. In the 16th century, Portugal and Spain established overseas colonies on a scale that had never before been witnessed. The economies of this core diversified and a wealthy merchant class emerged, which would over time provide the necessary capital required for industrialisation to occur. Consequently, later stages of development witnessed shifting emphasis to industrial production in Europe and the search for new markets. For example, by 1860 half of Brazil's imports were sourced in the UK. When South America, Asia and Africa entered the world system in the 19th century they were peripheral zones but investment in the peripheral and semi-peripheral zones during the 20th century encouraged the development of industry. During the 1950s and 60s, many Latin American countries pursued **import substitution policies** to facilitate rapid industrialisation and the restructuring of their economies. This involved replacing imports with domestically-produced goods.

Trade between the world's emerging markets has almost doubled since 2000, with trade involving a developed country accounting for the vast majority of total global trade, although the proportion has been decreasing. World exports represent about a third of global GDP. Consequently, international trade remains the fastest way for developing countries on the periphery to prosper and increase their GDP per capita.

FIGURE 4.17 Wallerstein's world regions

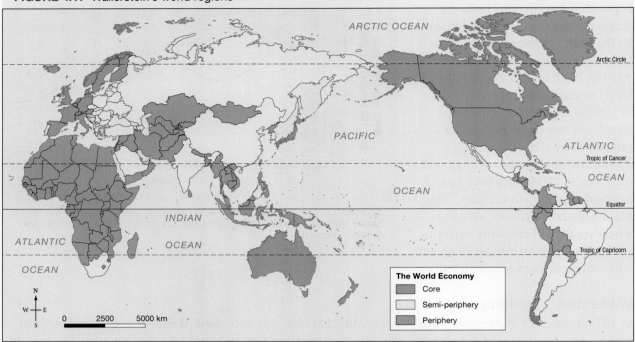

Source: Erin H. Fouberg, Alexander B. Murphy, Harm J. de Blij, Figure 8.10, Human Geography: People, Place, and Culture, 9th Edition, John Wiley & Sons.

FIGURE 4.18 Population growth of selected countries and regions, 1500–2000

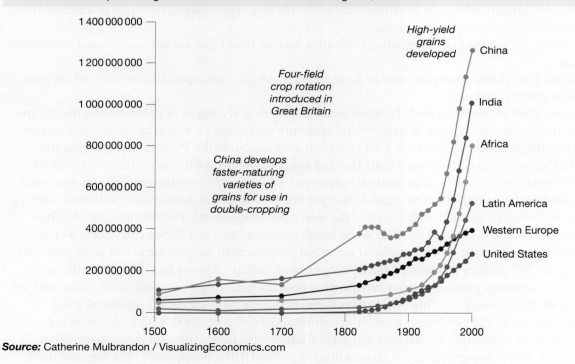

Source: Catherine Mulbrandon / VisualizingEconomics.com

Curvilinear relationships are variable, more complex and less easily identified than simple linear relationships, where the ratio of change is constant. At the higher ends of each variable in a curvilinear relationship there is little variation, whereas at the lower end there is more variability.

198 Jacaranda Senior Geography for Queensland 2 Units 3 & 4 Third Edition

Activity 4.2c: Analysing flow diagrams and correlations

Comprehend and analyse the patterns of population growth and economic development

1. (a) With the aid of Excel and using data from table 4.2, examine the correlation between population growth and economic development for the sample of 20 countries by constructing a scattergraph.
   (b) Refer to the information on curvilinear relationships in the previous box and interpret your result.

TABLE 4.2 GDP per capita versus total fertility rate

Country	GDP per capita ($US PPP), 2018	Total fertility rate, 2017
Australia	48 460	1.8
Bolivia	7559	2.9
Brazil	15 483	1.7
Burundi	770	5.7
China	16 806	1.7
Czech Republic	36 327	1.6
Ecuador	11 617	2.5
France	42 850	2.0
Ghana	4641	4.0
India	7055	2.3
Indonesia	12 283	2.4
Liberia	826	4.6
Malaysia	29 431	2.0
Netherlands	52 503	1.7
Niger	1016	7.2
PNG	4197	3.7
Thailand	17 870	1.5
Turkey	26 502	2.1
UK	43 268	1.8
USA	59 531	1.8

Source: © 2019 The World Bank Group

Analyse the relevance of Wallerstein's theory

2. Compare figures 4.9 and 4.17. How far do you think Wallerstein's concept of core periphery remains relevant?
3. (a) Study figure 4.18. Compare the pattern of population growth during the period 1500–2000 between:
   i. Western Europe and the USA
   ii. Latin America, India and China.
   (b) To what extent do you think the differences between the core and the periphery support Wallerstein's model of sequential development?

> **Resources**
>
> **Video eLesson** SkillBuilder: Constructing and interpreting a scattergraph (eles-1756)
> **Interactivity** SkillBuilder: Constructing and interpreting a scattergraph (int-3374)
> **Weblink** Compare the Rostow and Wallerstein models

4.2.5 The demographic transition model

In the early 20th century Warren Thompson, an American demographer, observed changes in population growth rates in the United States and other countries for which census data was available. He subsequently used the birth and death rate data to devise a threefold classification of countries according to their rates of natural change. He grouped countries into three categories:

- Countries with high birth and death rates
- Countries with declining and therefore lower death rates
- Countries with low birth and death rates.

According to Thompson, countries transitioned from having high birth and death rates to low birth and death rates as they became increasingly industrialised and democratic.

In the 1950s, another American demographer, Frank Notestein, was credited with refining Thompson's theory of population transition. The **demographic transition** model, as it is now known, has four distinct phases. Instead of three groups of countries, Notestein identified four by splitting the middle group into an early expanding segment where the birth rate remained high but the death rate declined, and a late expanding segment where the death rate was low but the birth rate declined. These are summarised in table 4.3 and shown in figure 4.19.

TABLE 4.3 Notestein's four stages of demographic transition

Stage 1	High birth rate	High variable death rate	Population remains stable
Stage 2	High birth rate	Declining death rate	Population begins to grow in size
Stage 3	Declining birth rate	Rate of death rate decline slackens	Population growth rate declines
Stage 4	Low variable birth rate	Low death rate	Population stability returns

Demographic transition in the UK

The British population increased slowly until around 1800. Prior to this, poor diet, famines, wars, and diseases restricted population growth. Both the birth and death rates were high, due to the bulk of the population living in rural areas, high rates of infant mortality and families needing to be self-sufficient to survive. Furthermore, the Black Death, which spread across Europe in the mid-14th century, is estimated to have wiped out half of the British population.

Industrialisation and **urbanisation** in the 1800s brought about important,

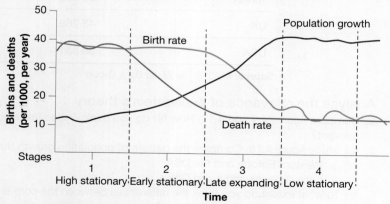

FIGURE 4.19 Notestein's demographic transition model

Source: © 2000–2019 All Rights Reserved Barcelona Field Studies Centre S.L.

although slow, changes to public health and food supply, which helped to bring down the death rate and increase life expectancy, the consequence of which was rapid population growth. However, falling birth rates, as a result of changing social attitudes and conditions such as the emancipation of women, led to a slackening of population growth. This decline was later furthered by access to contraceptives and the changing role of women in a previously male-oriented workforce. At the same time, the development of medicines, such as antibiotics to combat disease, helped to increase life expectancy. Despite birth and death rates remaining low in recent decades, the UK population has continued to grow, largely through **net migration**. In 2017 the total population of the UK reached 66 million (see figure 4.20).

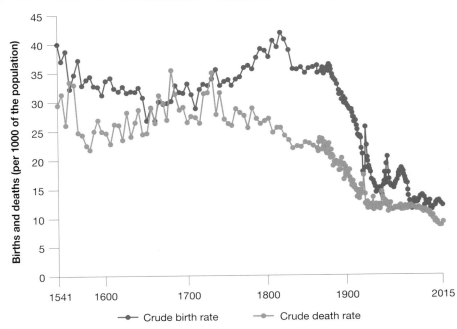

FIGURE 4.20 The demographic transition in England and Wales

Source: Our World In Data

The four-stage transition model has been a useful tool for analysing patterns of population dynamics. However, the birth rate of some countries in the developed world, particularly in Europe, have become so low that natural change is negative and unless they have an influx of migrants their populations decline. It has been suggested that these countries are perhaps indicative of a fifth stage in the transition model. Germany's national statistics office estimates that the country's total population will shrink from 83 million in 2018 to 73 million by 2060. This is despite its current annual influx of more than 100 000 migrants. Germany's rate of natural change has been in negative territory since the 1970s.

Rostow's stages of growth and Wallerstein's world systems theory are early theories about economic development and population with a focus on the economic system. The demographic transition model then used demographic data to classify countries. However, this model doesn't adequately account for the shrinking population in some developed countries. These models represent early thinking about population and more contemporary models will be discussed in section 4.5.4.

Resources

Weblink Demographic transition model
Video eLesson SkillBuilder: Creating a simple column graph (eles-1639)
Interactivity SkillBuilder: Creating a simple column graph (int-3135)

4.3 An ageing world

4.3.1 Declining fertility and increasing life expectancy

Many parts of the world have witnessed increasing life expectancy and falling fertility rates, which has led to an ageing population. This is when there is an increasing proportion of older people in a population. The UN has estimated that by 2050 the proportion of the world's population aged 60 and over will be equal to that of the population aged under 15 (see table 4.4). This translates into every major region in the world, except Africa, having at least a quarter of its population aged 60 and over. Europe has already reached this figure. In terms of raw numbers, the world population of this age group will increase from 962 million in 2018 to just over 2 billion in 2050.

The older population is growing at a faster rate in urban areas than in rural areas. Globally, between 2000 and 2018, the number of people aged 60 years and over increased by nearly 70 per cent in urban areas, whereas the increase in rural areas was only 25 per cent. This means that older persons are increasingly concentrated in urban areas.

TABLE 4.4 Per cent of population 60 years of age and over

Macro region	2015	2050
Africa	5.4	8.9
Asia	11.6	24.6
Europe	23.9	34.2
South America and the Caribbean	11.2	25.5
Oceania	16.5	23.3
North America	20.8	28.3
World	12.3	21.5

Source: World Population Ageing Report 2015 by United Nations / Population Division. Copyright © 2015 United Nations. Used with permission of the United Nations.

Activity 4.3a: Interpreting data about age–sex structure using tables, and column and scatter graphs

1. Using data from table 4.4, construct double column graphs for the regions listed.

Analyse the data and apply your understanding

2. (a) Describe the regional differences in 2015 revealed by your graph.
   (b) How is the regional pattern expected to change by 2050 in terms of differences in growth between the regions and the rank order?
   (c) Why do you think world regional differences will remain in 2050?
3. (a) Using data from table 4.5, construct a scattergraph on semi-log paper to show the relationship between the projected change from 2017 to 2050 in the number of persons aged 60 years and over, and gross national income per capita in 2016.
   (b) Interpret your result and explain what it tells us about the relationship between the expected growth of the 60 and over age group and economic development during the next 30 years.
   (c) What do the growth figures suggest about the developing world's share of the global population aged 60 and over in the future?
   (d) What sort of challenges do you think this growth presents to the countries faced with an ageing population?

TABLE 4.5 Population aged 60 and over versus GDP per capita

Country	Predicted percentage change in population aged 60 and over 2017–2050	Gross domestic product per capita ($US 000) 2017
Rwanda	275	0.8
Tanzania	272	1.1

(continued)

TABLE 4.5 Population aged 60 and over versus GDP per capita (continued)

Country	Predicted percentage change in population aged 60 and over 2017–2050	Gross domestic product per capita ($US 000) 2017
The Gambia	265	0.5
Malawi	306	0.3
Zimbabwe	309	1.3
Cameroon	246	1.6
Ghana	227	1.7
South Africa	144	6.6
Egypt	208	2.6
Indonesia	171	4.0
India	153	2.1
China	109	10.1
Puerto Rico	53	31.3
Jamaica	99	5.3
Brazil	160	10.2
Uruguay	49	18.1
Australia	83	59.6
USA	55	62.5
UK	50	44.1

Source: International Monetary Fund 2017.

Semi-log graph paper has one axis that is arithmetic and one that is logarithmic. The arithmetic axis runs in 10s, increasing in increments of one. For example: 1, 2, 3, 4, 5, 6, 7, 8, 9, 10, then 11, 12, 13 and so on. On the log axis, each cycle runs linearly in 10s but the increase from one cycle to another is an increase by a factor of 10. For example: 1, 2, 3, 4, 5, 6, 7, 8, 9, 10, then 20, 30, 40 and so on.

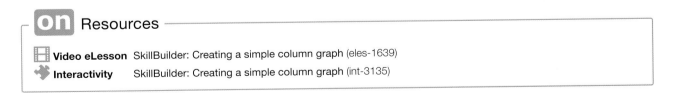

Resources

Video eLesson SkillBuilder: Creating a simple column graph (eles-1639)
Interactivity SkillBuilder: Creating a simple column graph (int-3135)

4.3.2 The challenges of an ageing population

Life expectancy is one of the major key indicators of development. Indeed, one of the components of the United Nations Development Program's human development index is life expectancy at birth. Consequently, an ageing population can be regarded as an indication of successful development. In 2016, life expectancy of the world population was 72 years, but there are large disparities between and across countries.

Japan has the highest proportion of people aged 65 and over in the world. This has been attributed to its traditionally low fat diet and the access its citizens have to a national healthcare program, which encourages regular medical examinations, especially for older generations. The traditional Japanese diet is low in calories and saturated fat but high in **nutrients**, especially phytonutrients, which are found in coloured vegetables. However, an ageing population presents challenges to governments.

The **dependency ratio** is the number of dependents in a population divided by the number of working-age people (see figure 4.21). The World Bank's definition of dependent population includes those people who are under 15 years of age and those aged over 64 even though it is recognised that many people in these two age groups work. This means that the higher the ratio, the greater the economic burden carried by working-age people. It has been estimated that in just one decade the cost of delivering aged care in Australia will double to $40 billion by 2028. Countries levy income and payroll taxes to help fund social security benefits, such as aged pensions, for those eligible to receive them. Some countries have responded to the increased cost of social services by raising the pension age. In Australia, the age you can receive a pension is 67 years, but many choose or find it necessary to continue employment beyond 67 years of age.

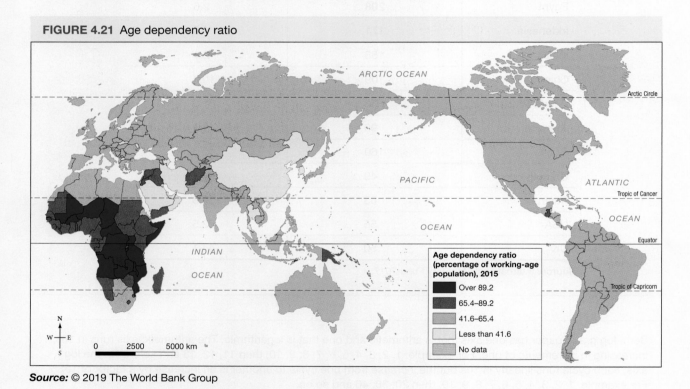

FIGURE 4.21 Age dependency ratio

Source: © 2019 The World Bank Group

Since 2000, Africa has recorded a significant improvement in life expectancy, despite having a lower average life expectancy than all other world regions. Improvements in health and welfare have made a significant contribution to this increase. Two of the biggest fatal diseases in Africa have been HIV/AIDS and malaria. The World Health Organization has predicted that HIV/AIDS could be eradicated by 2030.

TABLE 4.6 Top five causes of death, Africa

1	Lower tract respiratory disease
2	HIV/AIDS
3	Diarrhoeal diseases
4	Malaria
5	Tuberculosis

Source: Based on data from World Health Organization

4.3.3 Ageing populations in China and Japan

Demographic transition in China

The People's Republic of China was founded in 1949. There is little reliable data available on the Chinese population before this time and, consequently, it is not possible to accurately determine when the death rate began to decline and when stage two of the demographic transition started. What we do know is that not long after the launching of The Great Leap Forward in 1958, the birth rate plummeted and the death rate spiked. The Great Leap Forward was an ambitious but short-lived project conceived by leader Mao Zedong to modernise the centrally planned Chinese economy. The aim was to boost both farm and industrial production by a series of major reforms using labour intensive methods. This involved the development of small backyard steel furnaces in every village and urban neighbourhoods, and the controversial forced movement of villagers into agricultural collectives or farming communes. Although initially successful, production failures such as in the grain harvest led to a widespread and devastating famine. The death rate spiked, especially among infants and the elderly.

The mortality rate returned to a declining pattern when food production improved. Further decreases can be attributed to social improvements, notably education, the expansion of primary health care services, which included the formation of so-called 'barefoot doctors' during the late 1960s, and childhood immunisation programs against diseases such as tuberculosis, polio and measles (see figure 4.23).

FIGURE 4.22 An aged Chinese farm labourer

Source: Michael Morrish

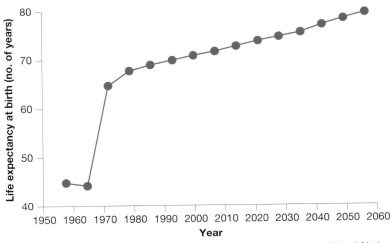

FIGURE 4.23 Life expectancy in China, 1950–2050

Source: World Population Prospects: The 2010 Revision, Volume I: Comprehensive Tables by United Nations / Population Division. Copyright © 2010 United Nations. Used with permission of the United Nations.

The one-child policy

China's one-child policy was established by Chinese leader Deng Xiaoping in 1979 to restrict population growth. As its name suggests, it was designed to limit couples to only one child. Public posters throughout China reminded its citizens of the government policy (see figure 4.24). Couples who did not comply faced fines and possible sterilisation of the female partner if found guilty of a subsequent pregnancy. However, the one-child policy only affected about a third of Chinese households, largely restricted to the ethnic Han Chinese living in urban areas.

FIGURE 4.24 (a) Public poster promoting the one-child policy in China and (b) Indian 2 rupees promoting small families

Source: (a) Michael Morrish (b) Paul Hayler

The policy was abandoned in 2015. Although credited with successfully curbing population growth, the country's fertility rate was already falling before the policy began. Furthermore, the policy has been held responsible for creating today's gender imbalance. Nationally, the male to female ratio in China is 106:100 but in the under 25 age group it is 114:100. Given the Chinese cultural preference for male heirs, the one-child policy was blamed for sex-selective abortions and even infanticide of female infants.

After 30 years of the one-child policy, China now faces the problems of a shrinking workforce and an increasingly ageing population. According to the Chinese National Bureau of Statistics, the total working population, aged between 15 and 64 years, has been falling since 2014. Meanwhile, the proportion of the population aged 65 and older has increased to more than 11 per cent. Increased life expectancy in China is partly responsible for growth in the elderly population. It is a phenomenon that is particularly acute in some cities and in the heavy industrial areas of China. The Shanghai Population and Family Planning Committee has predicted that by 2020 more than a third of Shanghai's population will be over 60.

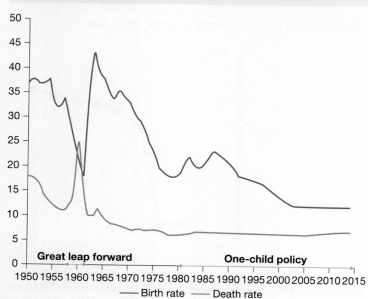

FIGURE 4.25 Birth and death rates in China, 1950–2015

China's dependency ratio for those aged 65 and older is rising. It was 14 per cent in 2015 and the UN has estimated that this could rise as high as 44 per cent by 2050. This ratio is important because it is an indicator of the number of dependents each person of working-age will, on average, need to support. Average life expectancy will have reached 80 years by 2050.

China's total fertility rate is currently lower than 2.1 births per woman, which is regarded as the replacement level of fertility. This means that the population will be declining unless there are more people migrating to China than leaving China. However, there is a limited quota for permanent residence permits.

Demographic transition in Japan

Japan has been experiencing a decline in population growth since the end of World War II. In the late 1940s, population growth dropped to replacement level of about 2.1 children per woman. This has been primarily attributed to declining rates of childbearing among married couples. Japan's fertility transition in the mid-1970s marked the beginning of absolute population decline associated with what some demographers believe is an additional fifth stage of the demographic transition model. The latest phase in Japanese population change has been associated with decreasing rates of marriage. 2017 not only marked the fifth consecutive year that fewer Japanese people married but the lowest number since 1948.

Delayed marriage appears to have resulted, in part, from employment conditions. As the number of women with higher education has grown so have their employment opportunities. At the same time, male employment rates have declined, which has impacted on men's marriage prospects. Coupled together, these two factors have contributed to the decline in the number of marriages and, therefore, the national fertility rate.

Japan has one of the highest rates of life expectancy in the world but, in 2018, the number of births recorded in Japan fell to its lowest level since records began in 1899 (see figure 4.26). This was the third consecutive year that births had been below one million. In contrast, the number of deaths increased in line with an ageing population over the one million mark.

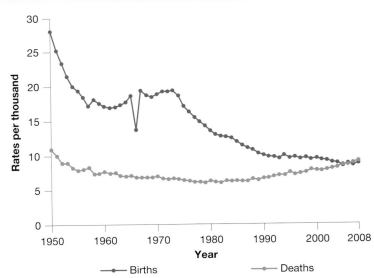

FIGURE 4.26 Birth and death rates in Japan, 1950–2008

The Japanese Ministry of Health, Labor and Welfare is examining ways to improve support for younger generations of married Japanese couples in employment. Japanese corporate culture has never made it easy for married women to achieve a good work-life balance, particularly surrounding maternity. Since the mid-1990s, the government has encouraged the provision of more childcare services but the number of children on waiting lists has grown faster than the supply and there are doubts that the target, a place for every child

by 2020, can be achieved. The problem is particularly acute in the large metropolitan areas. The income-tested child allowance scheme has been broadened but remains small compared with those offered in other countries. The income compensation scheme introduced in 1995 is considered too low and a large number of employers, especially smaller organisations, have never formulated any policy for parental leave. The Japanese government has sought a number of solutions which include:
- making access to childcare easier for women, to allow them to return to work after their child is born.
- raising GST by 2 per cent to help offset the spiraling cost of its social security and health services.
- considering a reduction in the tax burden for part-time employees and making interest-free loans available for higher education.

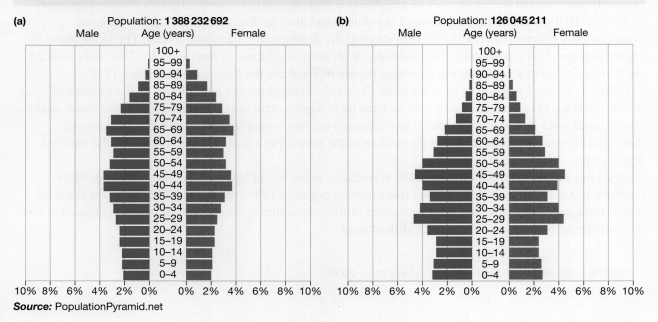

FIGURE 4.27 Population pyramids for (a) Japan and (b) China, 2017

Source: PopulationPyramid.net

Activity 4.3b: Analysing demographic transition in China and Japan

Explain and analyse the data

1. (a) Refer to figure 4.27 and compare the shapes of the population pyramids for Japan and China.
   (b) What do the two pyramids tell us about each country's future age structure?
2. Study the population pyramids A to E shown in figure 4.28, paying particular attention to their different shapes, heights and bases. For each pyramid, state the stage of the demographic transition it most likely represents and explain your answer.

Apply your understanding using graphs

3. (a) Using a copy of figure 4.25, and with the aid of figure 4.19, insert two vertical lines to identify stages 2, 3 and 4 of the demographic transition.
   (b) Justify the positioning of your lines.
   (c) What is unusual about China's birth and death rates in stage 2 compared to the model?
   (d) Explain this anomaly.
4. (a) Using data from table 4.7, construct a double line graph to show the total fertility and population growth rates for China between 1950 and 2015.
   (b) Insert a horizontal line to show the replacement rate of 2.1.
   (c) Insert two vertical lines to indicate the beginning and end of the one-child policy.
   (d) Compare the patterns of change during this period.
   (e) Explain the link between the two rates.

FIGURE 4.28 Population pyramids

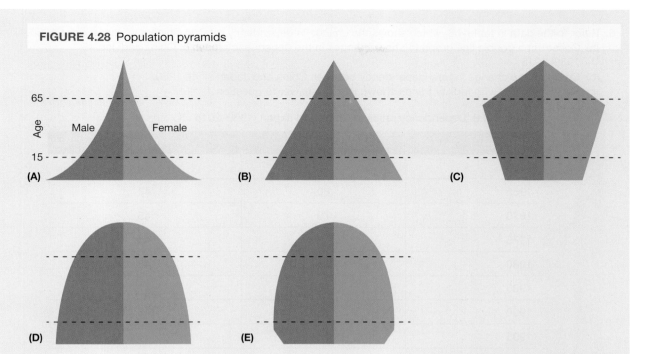

5. (a) Using data from table 4.7, construct a double line graph for Japan.
   (b) Compare the changes in total fertility and population growth between China and Japan.
   (c) Explain the similarities and/or differences between the two countries.

TABLE 4.7 Fertility rates and population growth in China and Japan, 1950–2015

	China		Japan	
Year	Fertility rate (%)	Population growth rate (%)	Fertility rate (%)	Population growth rate (%)
1950	6.0	2.0	3.5	+1.8
1955	5.4	1.9	2.5	+1.1
1960	6.2	1.8	2.0	+0.9
1965	6.2	2.4	2.1	+0.9
1970	4.7	2.7	2.1	+1.1
1975	3.0	1.8	1.9	+1.6
1980	2.5	1.3	1.7	+0.8
1985	2.7	1.4	1.8	+0.6
1990	1.9	1.5	1.5	+0.3
1995	1.5	1.1	1.4	+0.4
2000	1.5	0.8	1.3	+0.2
2005	1.6	0.6	1.3	+0.0
2010	1.6	0.5	1.4	+0.0
2015	1.6	0.5	1.4	−0.1

Source: United Nations Statistics Division, 2017.

6. Refer to the data in table 4.8, which shows the change in dependency rates in China and Japan since 1960.
   (a) Construct a double line graph to show changes in the dependency ratios of China and Japan 1960–2015.
   (b) Compare the change in total dependency between China and Japan since 1960.
   (c) Explain the link with fertility trends shown in your answer to question 4.

TABLE 4.8 Dependency ratios in China and Japan, 1960–2015

Year	China	Japan
1960	76	56
1965	80	47
1970	79	45
1975	78	47
1980	68	48
1985	56	47
1990	52	43
1995	51	44
2000	46	47
2005	38	50
2010	36	56
2015	38	64

Source: © 2019 The World Bank Group

Resources

Video eLessons SkillBuilder: Constructing multiple line and cumulative line graphs (eles-1740)
Skillbuilder: Comparing population profiles (eles-1704)

Interactivities SkillBuilder: Constructing multiple line and cumulative line graphs (int-3358)
Skillbuilder: Comparing population profiles (int-3284)

4.4 Patterns of changing population distribution and density

4.4.1 Factors influencing population distribution

Over time, physical and socioeconomic factors have influenced both the distribution and density of the world's population. At one end of the spectrum are the densely populated regions that include the highly urbanised and industrial regions of north-west Europe, north-east US and Canada, and the agricultural heartland of Asia. At the other extreme lies the high latitude and high altitude regions, which include the frozen polar north and continent of Antarctica, the mountainous belts of the Andes, Rockies and Himalayas, and the hot desert zones of Saharan Africa and central Australia.

The staggering growth of the global population has been accompanied by significant shifts in its geographical distribution. According to the UN, developing countries accounted for about two-thirds of the world's population in 1950, but by 2030 this figure is expected to have increased to 85 per cent. Developing countries having contained more than 95 per cent of the world's population growth since 2000 (see figures 4.29 and 4.30).

FIGURE 4.29 Distribution of the global population by continent

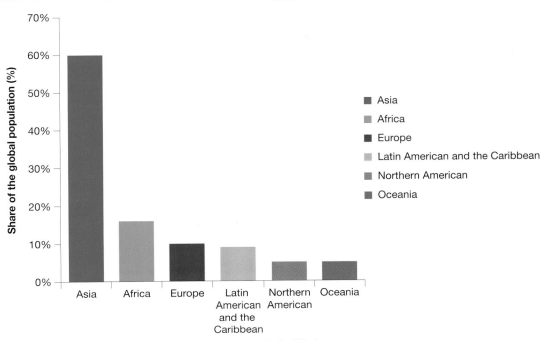

Source: Created by John Wiley & Sons based on statistics from United Nations

4.4.2 Impact of changing population distributions and densities

Population distribution around the world is changing. Although Asia is by far the most populous continent, Africa is the fastest growing region in the world and its share of global population is expected to rise to 20 per cent by 2050. The population of sub-Saharan Africa alone is predicted to double to more than 2 billion by 2050. In contrast, Europe's share of world population is expected to fall from about 20 per cent in 1960 to less than 10 per cent by 2050.

Demographic shifts in population distribution will impact different parts of the world in different ways. Shifts in population distribution also result in changes in population density given that population density is a measure of population distribution in a particular geographic area. For example, the increasing concentration of the world's population in towns and cities has aggravated existing problems, such as inadequate housing, and poor air and water quality. In sub-Saharan Africa, one of the most pressing challenges will be reducing **food insecurity**, particularly in cereals, and reliance on imported food. This will require increased productivity, which involves closing the gap between current farm yields and **yield potential**.

Rainfed maize is the dominant cereal in sub-Saharan Africa. It has the greatest yield potential but the largest **yield gap**. Millet, another cereal staple in the region, has the smallest yield potential and the smallest yield gap. Sub-Saharan Africa needs a green revolution but Food and Agriculture Organization (FAO) economists say that the chances of this are slim. To achieve the same success as other parts of the world, such as Asia, there needs to be financial incentives for the necessary genetic engineering, improvements in water supply and a substantial reduction in the extent of the region's degraded soils.

4.4.3 Distribution of population within countries

Although there are global trends in population distribution for different regions, within world regions, individual countries can also have patterns and variations in population distribution. The distribution and density of population within countries also reflects the physical and socioeconomic factors that influence world population. For example, in China there is a major demographic divide between the west and the east. Provinces to east of the Heihe–Tengchong line account for a third of the country's land area but are home to more than 90 per cent of the country's total population. Unequal population distribution is also evident in Cambodia.

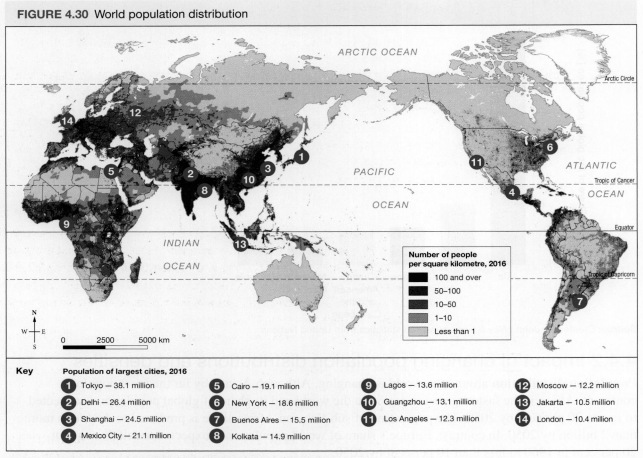

FIGURE 4.30 World population distribution

Source: The Earth Institute Columbia University / United Nations

Cambodia: Measuring unequal distributions

Located in south-east Asia (see figure 4.31), Cambodia was once part of the French colony of Indochina and gained independence in 1953. Bisected by the Mekong River, Cambodia is physically dominated by its extensive and fertile central lowlands, on which two-thirds of the country's 16 million inhabitants live.

The distribution of the Cambodian population changed dramatically during the 1970s when a mass genocide was carried out under the rule of the Khmer Rouge, led by the dictator Pol Pot. Large numbers of people fled the fighting in the countryside to seek refuge in the cities, only to be forcibly relocated back to rural areas along with city dwellers. Vietnam invaded Cambodia in 1978 to depose Pol Pot and, since the end of this war in 1991, there has been further redistribution of the population. The newly elected government in 1993 embarked on a program of economic reform and improvements in infrastructure to provide greater connectivity between the provinces. Consequently, the proportion of the workforce employment in agriculture fell from three-quarters in 1993 to two-thirds in 2010. The cities of Phnom Penh and Kampong Cham, and the Takeo province have more industrial enterprises than others, and this correlates with their pattern of higher provincial **population density**. The provinces with low population density, such as Kratie and Stung Treng, all have the fewest number of industrial enterprises.

FIGURE 4.31 Cambodia

Source: Made with Natural Earth. Free vector and raster map data @ naturalearthdata.com.

THE LORENZ CURVE

The Lorenz Curve was developed by Max Lorenz to measure the degree of inequality of wealth within a country. It has proved useful in fields such as geography and ecology, to identify uneven distributions of plant species.

The 45° line indicates a perfectly even distribution. All the points along the line are of equal percentages. For example, 40 per cent of the population inhabits 40 per cent of the land area. Therefore, the further away the Lorenz Curve is from the 45° line, the more uneven the distribution.

FIGURE 4.32 Lorenz Curve for a population distribution

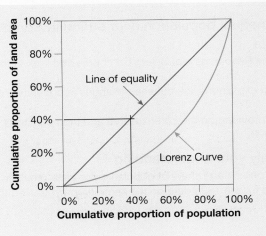

CHAPTER 4 Global population change 213

HOW TO CONSTRUCT A POPULATION DISTRIBUTION LORENZ CURVE

1. Rank the census districts in declining order. The district with the largest population is rank 1, the second largest is rank 2 and so on.
2. Calculate the percentage of people in each district.
3. Calculate the cumulative percentage by adding successive percentages from rank 2 onwards. In the following example the cumulative percentage for rank 2 is 40 + 30 = 70 per cent.

Census district	Rank	Number of people	Percentage of total population	Cumulative percentage
A	1	800	40	40
B	2	600	30	70
C	3	300	15	85
D	4	200	10	95
E	5	100	5	100
		2000		

4. Plot the cumulative data on the graph against the rank order and join the points.
5. Interpret the graph. The further the Lorenz Curve is from the 45° line (when both axes are of equal length), the more uneven the distribution of population.

FIGURE 4.33 Plotting data on a Lorenz Curve graph

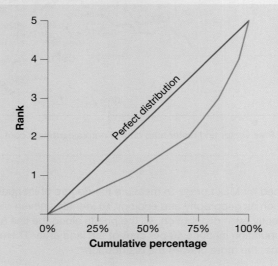

Activity 4.4: Measuring unequal population distribution in Cambodia

Synthesise and apply data

1. (a) Access **Gap Minder** in the Resources tab. Select Cambodia and look at the trends icon and income per person (GDP).
   (b) How does this graph help explain the link between changes in population distribution after the conflict with Vietnam and GDP?
2. Read the boxed information provided on the Lorenz Curve and then construct the Lorenz Curve for Cambodia's population in 2008.

Analyse the data and apply your understanding

3. (a) Calculate the cumulative percentage share of population for each province in 1998.
   (b) Construct a Lorenz Curve for the 1998 data on the same graph you used for question 1.

(c) How does the distribution in 2008 compare with that of 1998?
(d) Explain why you think the distribution has changed.

TABLE 4.9 Cambodia's population distribution

Province/Municipality	Population, 2008	Share of total (%)	Land area (km²)	Share of total (%)	Population, 1998
Mondul Kiri	61 107	0.46	14 288	8.03	28 576
Stung Treng	111 671	0.83	11 092	6.23	77 644
Koh Kong	117 481	0.88	10 090	5.67	121 080
Preah Vihear	171 139	1.28	13 788	7.74	124 092
Ratanak Kiri	150 466	1.12	10 782	6.06	97 038
Kratie	319 217	2.38	11 094	6.23	266 256
Otdar Meanchey	185 819	1.39	6158	3.46	67 738
Pursat	397 161	2.96	12 692	7.13	355 376
Kampong Thom	631 409	4.71	13 184	7.76	540 544
Kampong Chhnang	472 341	3.53	5521	3.10	419 556
Siem Reap	896 443	6.69	10 299	5.78	700 332
Battambang	1 025 174	7.65	11 702	6.57	795 736
Pailin	70 486	0.53	803	0.45	23 287
Banteay Meanchey	677 872	5.06	6679	3.75	581 073
Kampong Speu	716 944	5.35	7017	3.94	596 445
Kep	35 753	0.27	336	0.19	28 560
Preah Sihanouk	221 396	1.65	1938	1.09	172 482
Kampot	585 850	4.37	4873	2.74	526 284
Svay Rieng	482 788	3.60	2966	1.67	477 526
Kampong Cham	1 679 992	12.54	9799	5.50	1 607 036
Prey Veng	947 372	7.07	4883	2.74	947 302
Takeo	844 906	6.31	3563	2.00	790 986
Kandal	1 265 280	9.45	3564	2.00	1 076 328
Phnom Penh	1 327 615	9.91	294	0.17	999 894
Total	13 395 682		178 035		11 421 161

Source: Ministry of Internal Affairs and Communications Statistics Bureau Japan https://www.stat.go.jp/info/meetings/cambodia/pdf/ci_fn02.pdf

Resources

🔗 **Weblink** Gap Minder

4.5 People on the move: international and internal migration

4.5.1 Why do people migrate?

Push and **pull factors** 'push' people to leave their own country or region and 'pull' people towards another, and help us to understand why people migrate. These factors can be examined in the context of **forced** and **voluntary migration** at both internal and international levels (see figure 4.34).

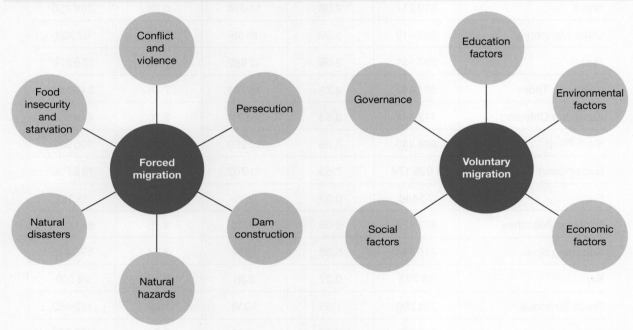

FIGURE 4.34 Reasons people migrate

Forced migration

Forced migrations are those where people have been forced into leaving an area against their will or as a result of life threatening circumstances. These include:
- natural hazards and disasters
- dam construction
- conflict
- persecution (religious or cultural)
- food and water insecurity.

Natural hazards and disasters

New Zealand's Christchurch earthquake of 2011 caused a population exodus of almost 10 000 people, mostly from the city's Red Zone, a 630 hectare abandoned residential area bordering the Avon River (see figure 4.35).

FIGURE 4.35 The Red Zone, Christchurch

216 Jacaranda Senior Geography for Queensland 2 Units 3 & 4 Third Edition

Liquefaction had caused so much damage in this zone that it required engineering solutions considered not to be cost-effective. The long-term safety and wellbeing of residents returning to the area was also considered to be at risk. The government purchased the area and nearly 8000 homes have subsequently been demolished. The area is now being redeveloped for recreation purposes and has been renamed the Okataro Avon River Corridor.

Climate change

The 2018 World Bank Report on **climate change** titled 'Groundswell: Preparing for internal climate migration', has predicted the creation of more than 143 million so-called climate migrants. These people will be fleeing sea level rise, water scarcity and crop failure. Most of this migration will take place in sub-Saharan Africa, south Asia, and Latin America, which are home to more than half of the developing world's populations.

The effects of **global warming** and sea level rises are particularly acute in the Pacific Islands region, where people are already migrating from island chains and low-lying coastal areas that flood regularly. For example, the Republic of Kiribati is made up of 32 scattered island atolls close to the equator, with most of the atolls less than three metres above sea level, making migration to higher ground not an option. Rather than wait for the inevitable, Kiribati is developing opportunities for its citizens to migrate to Australia and New Zealand. The Migration with Dignity Policy, as it is known, partly subsidises relocation costs and provides educational and vocational opportunities for islanders to improve their employment prospects in their new homes. However, those with limited skills are likely to be left behind until they are forced to move.

Dam construction

Dam construction, due to its large scale and the **resources** required, has the ability to displace millions of people. It estimated that China has resettled more than 20 million people since 1949 to make way for water storage reservoirs and hydropower projects. The massive HEP Three Gorges Dam Project on the Yangtze River alone has resulted in the forced resettlement more than 1.2 million people since 1993. More recently, China has embarked on an ambitious plan to increase its hydro-electric power capacity by constructing dams on the Mekong River in the country's southwest.

The Mekong River delta, located in Vietnam, is one of the most vulnerable deltas in the world to sea level rise. Mekong Delta is predominantly covered by low-lying floodplains measuring 0.5 to 3 m above sea level. However, the construction of so many dams has resulted in a decrease in freshwater and sediment delivery to the delta. Consequently, the combined effect of the rising sea level and salinity intrusion on agriculture is threatening the economy and livelihood of local populations. Vietnam's Land Law gives the government the right to acquire land on the basis of protection against natural disasters and relocate residents to areas considered safer.

Persecution

The migration of refugees is an example of forced migration (see figure 4.37). The United Nations defines a refugee as a person who has fled their own country and has a well-founded fear of persecution due to their race, religion, nationality, social group or political opinion, and for whom it is therefore deemed unsafe to return. It is estimated that refugees now make up 10 per cent of all international migrants. An **asylum seeker** is similar, having fled their own country due to fear their life is at risk but has not yet had their claim for protection in another country assessed.

In 1972 the president of Uganda expelled the country's Asian population as part of his Africanization Program, with little warning. Descendants of largely Indian migrants, who had been encouraged to settle in Uganda in the early 20th century to build the East African Railway, had become a successful commercial elite in the country. This made them unpopular with the military dictatorship that came to power in 1971, which falsely accused them of exploiting the Ugandan economy at the expense of the indigenous Ugandans. By the end of 1972, the population of 96 000 Asian people had shrunk to barely 1000. Many ended up destitute in England or in whatever country would take them.

FIGURE 4.36 Asylum seekers crossing the Aegean Sea from Turkey to Greece

Conflict

More than 1.3 million asylum seekers entered the European Union (EU) in 2015 (see table 4.10). Most came from conflict-ravaged Syria, as well as Afghanistan and Iraq (see figure 4.38). They reached the EU via Turkey, having crossed Balkan countries, such as Serbia and Macedonia, by foot. However, another 100 000 asylum seekers migrated from sub-Saharan African countries, reaching Italy and Greece by making perilous crossings of the Mediterranean Sea from North Africa. Many migrants used these seaboard countries as a stepping stone for their preferred destination, Germany.

The sheer volume of asylum seekers took many EU governments by surprise and resulted in different responses to the crisis. Germany adopted an open door policy, accepting nearly 900 000 people, and the city of Mannheim became the terminus for special trains carrying refugees travelling via the Balkan route until it was closed in 2015. Three former sprawling US army barracks were converted into emergency accommodation to help house 12 000 refugees.

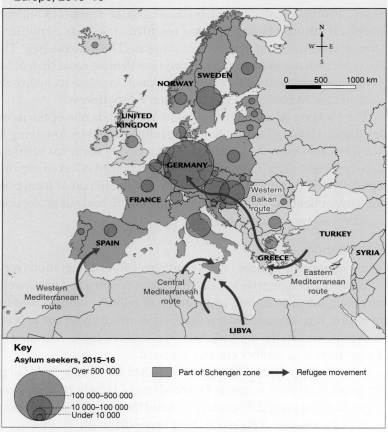

FIGURE 4.37 Movement and destination of asylum seekers into Europe, 2015–16

Source: Eurasian Research Institute, UNHCR, Eurostat

TABLE 4.10 Number of asylum applications in the EU, 1985–2018

Year	Number of asylum applications to EU ('000)
1985	159.1
1990	397.0
1995	263.6
2000	405.2
2005	227.5
2010	259.4
2015	1332.8
2018	638.2

Source: European Commission, Eurostat

Landings by asylum seekers peaked in 2016 as containment measures began to take effect. The enforcement of registration in Italy helped stem the flow of migrants northwards. The migration crisis had already become a housing crisis, with informal camps established across Italy and in key border crossings such as Calais in France. This camp, which housed more than 10 000 people, was demolished by French authorities in late 2016 and the residents were forcibly removed to other reception centres across the country.

FIGURE 4.38 Movements of displaced Syrians, 2017

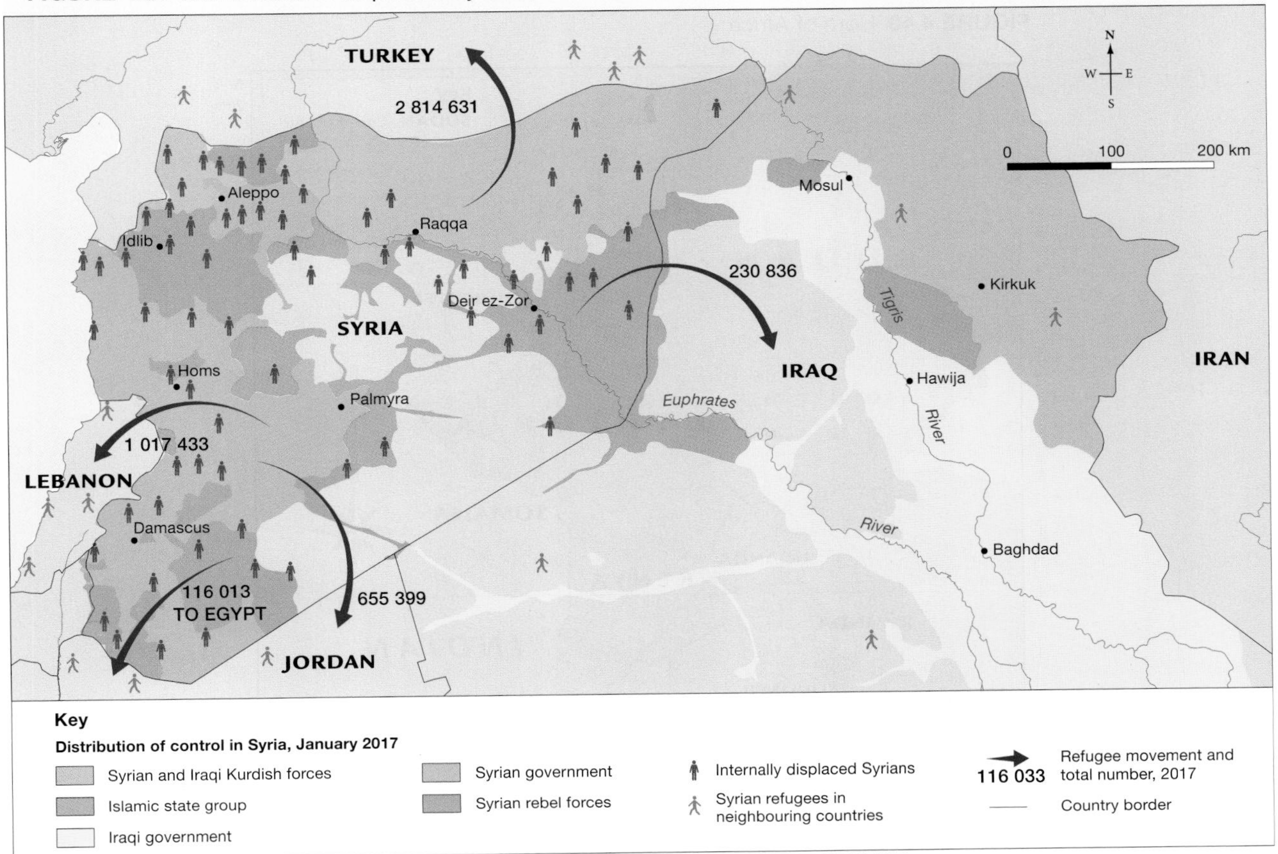

Source: Redraw by spatial vision based on the information from IHS Conflict Monitor and UNHCR

FIGURE 4.39 Syrians fleeing airstrikes and artillery shelling

Source: EPA / United Nation Relief and Works Agency / AAP Newswire

Drought and food insecurity

The Greater Horn of Africa (see figure 4.40) has been identified by the World Bank as a hotspot for climate migrants. The increasingly variable rainfall patterns have affected the region's agricultural production systems, resulting in acute food insecurity episodes since 2015. The unstable weather conditions have been brought about by both an El Niño event, which caused suppressed rains in the northern sector and floods in the equatorial sector of the region, and then a La Niña event, which triggered droughts in the equatorial part of the region. Conflict in the Sudan has only served to worsen the food insecurity.

FIGURE 4.40 Horn of Africa

Source: Natural Earth Data

Migration due to famine tends to be short-term and temporary, and is the last option left to people at the risk of starvation. In 2016, it is estimated there were more than one million internally displaced persons in Somalia. The majority were from the south and central regions, where about 25 per cent of the total population are classified as displaced. Drought and food insecurity were identified as the main push factors, along with tribal clashes.

Voluntary migration

Voluntary migration occurs when a migrant has a choice as to whether or not they migrate to a particular place. There are several factors that influence a migrant's decision-making and they often do not operate in isolation. Instead, a combination of factors is likely to influence a migrant's decision to move. Pull factors that encourage voluntary migration at internal and international level include:
- economic
- education
- social
- environmental
- governance.

Economic factors

Migrants are often attracted to destinations that offer better employment opportunities and high wages. However, high levels of income and other taxes may distort the flow of migrants despite the better employment opportunities.

Large-scale Polish migration to the UK dates from World War II, as a result of Polish servicemen becoming part of the Allied forces, and the establishment of the Polish government-in-exile in London following the invasion of Poland by Germany in 1939. After the war, these Polish people were resettled in the UK under the Polish Resettlement Act of 1948 and encouraged to move into key areas of labour shortage. This laid the foundations for Polish communities in towns and cities across the country.

The Polish People's Republic, a socialist regime, existed from 1947 to 1989, and placed mobility restrictions on the Polish people. When these restrictions were lifted in 1989, there was an increase in Polish migration to the UK. When Poland joined the EU in 2004, their citizens gained the right to move to the UK to live and work. This event resulted in much larger numbers coming to the UK, where the economy was booming. This is an example of what happens when there is an imbalance between labour demand and supply, and how migration can rectify this imbalance.

A large discrepancy between unemployment rates in the UK and Poland led to a large number of Poles applying for work in the UK. In 2005, the UK unemployment rate was 5.1 per cent compared with Poland's 18.5 per cent. Job vacancies in the UK were also particularly high, in excess of 600 000. The minimum wage a migrant could earn at the time was also twice as high as the average Polish worker's take-home pay. By 2007, more than 350 000 Poles had applied for work in the UK. According to the 2011 UK census, the Polish community numbered 579 000, making it the second largest ethnic group in the country. A large majority of the migrants were employed in the UK industrial, warehousing, packing, catering, cleaning and farming sectors.

Unlike many other ethnic minorities, the settlement of Poles in the UK has been dispersed. This has resulted in most towns and cities possessing a Polish presence whether it is in the form of grocery stores and restaurants selling Polish food (see figure 4.41), or English pubs selling Polish beer.

In 2016, the UK referendum, the so-called Brexit Vote, asked citizens to decide whether or not to remain in the European Union and saw a narrow margin of voters in favour of the UK leaving. The impact on migration was immediate and a surge in the number of Polish people and other EU citizens leaving the UK resulted in a sharp fall in net migration to the UK, to under 250 000. This was the lowest level for nearly three years. This decline is likely to continue as a result of the economic uncertainty surrounding the Brexit Vote and the subsequent bitter dispute between the British Government and the European Union.

Education factors

Students are the fastest growing group of international migrants. According to the Organisation for Co-operation and Development (OECD), the growth in the number of students enrolled in tertiary education outside their country of citizenship has grown on average by 7 per cent each year since 2000. Many universities across the world have been keen to attract students from overseas because they are an important source of revenue. This phenomenon is known as globalisation of education.

FIGURE 4.41 Polish grocery store in the UK

In the UK, student migration has become a significant share of immigration, having grown steadily since the 1990s. In 2017, an estimated 184 000 higher education students were attending UK colleges and universities. Seventy per cent were from non-EU countries, with most of these students coming from China, India and the United States.

Social factors

Social factors such as friendships and family reunions can be significant in determining where a migrant chooses to live, particularly within the host country. Migrants are more likely to settle in their new home when they are surrounded by people of similar culture, beliefs and language. This can lead to the creation of ethnic enclaves (see section 4.5.3).

Environmental factors

In the developed world, positive socio-environmental factors have influenced internal migration patterns, usually in tandem with economic prosperity. Lifestyle, lower costs of living and good employment prospects have been significant factors in the high volume of inter-state migrants to Queensland since the 1960s.

In the US, roughly 14 per cent of the population is on the move every year. Although much of this involves relocating within cities, Americans moving to the warmer climates of the so-called Sun Belt have been shaping migration patterns in recent decades. The Sun Belt stretches across the southern and southwestern portions of the country from Florida to California. It includes states such as Texas, where migration accounted for half of its population growth between 2010 and 2016. This suggests that Texas and most other Sun Belt states will probably experience the growth levels they achieved prior to the recession of 2006–9. However, California is a notable Sun Belt exception. Although fewer migrants left California than other Sun Belt states during the recession, people are now choosing to move to more affordable states such as Nevada, Arizona, and Oregon as their economies grow.

Governance

Governments can play a pivotal role in both internal and international migration. The origins of free movement of citizens within the EU date back to the Treaty of Rome in 1957, which established what was then called the European Economic Community (EEC). Known as a customs union, the principal aim was to encourage free trade between member states, but have the protection of a common external tariff barrier. Originally comprising just six members, France, Germany, Italy, Belgium, the Netherlands and Luxembourg, the ultimate goal was the free movement of not just goods but also services, capital, and labour between them.

4.5.2 International migration

Currently, there are more than 250 million people living in a country other than their country of birth. The 2018 United Nations International Migration Report estimated that international migrants as a proportion of the world's population had increased from 2.8 per cent in 2000 to 3.4 per cent. Moreover, most of the growth in the global population of international migrants has been caused by movements toward high-income countries. This trend is expected to continue as a result of increasing connectivity, continued regional inequalities and demographic imbalances.

The 2030 Agenda for Sustainable Development set out 17 goals to end global poverty, fight climate change, encourage world peace, and strengthen global bonds. To help achieve these objectives, UN member states committed themselves to protecting the human rights of all migrants, and countering xenophobia and intolerance directed towards migrants. They also agreed to share the burden of hosting and supporting the world's refugees more equitably. The Global Compact for Migration, as it is known, is the first-ever UN global agreement on a common approach to international migration.

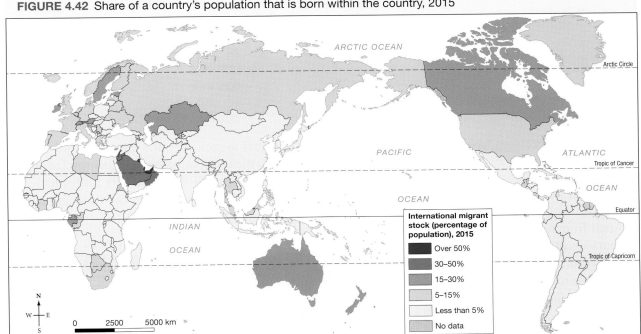

FIGURE 4.42 Share of a country's population that is born within the country, 2015

Source: © 2019 The World Bank Group

4.5.3 Impact of migration

Migration impacts on both the host area to which people move and the area from which they have moved. It is age selective, which means that, all things being equal, the more youthful elements of a population are more likely to migrate than the aged. The median age of migrants who move from one country to another is 39 years, but the age varies across the world. The average age of immigrants into the EU is 27 years (see figure 4.43). Consequently, migrants can help to reduce the age dependency ratio in countries that have an ageing population. They can, as witnessed in parts of Europe, even reverse population decline when the fertility rate falls below the replacement level.

Population growth as a result of net migration gains can have an economic impact. Like everyone else, migrants are consumers and will increase the total spending on goods and services within an economy. This in turn should lead to an increase in real GDP. By being gainfully employed, migrants will also pay income tax, GST on the goods they purchase and are probably not eligible for welfare benefits unless the size of their families is above the minimum threshold for such financial support.

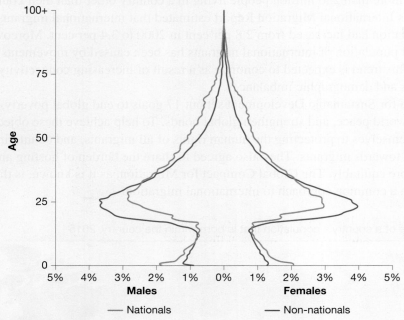

FIGURE 4.43 Age structure of migrants to the EU, 2016

Source: European Commission, Eurostat

Acute skills shortages have often prompted national governments to seek potential migrants from other countries. In the 1950s, Britain was short of workers within the transport system, postal service and hospitals. Former British colonies in the West Indies, such as Trinidad and Tobago, became a reliable source of labour to meet shortages. In 2018, over a third of the UK's registered doctors had been trained overseas.

Apart from reducing unemployment in countries of origin, it is not unusual for migrants working abroad to send remittances to the families they have left behind, particularly if they view their relocation as only short-term. The importance of migrant remittances as an external source of capital for developing countries has been recognised by the International Monetary Fund (IMF) since the 1980s. They are the second largest source of external financing in developing countries after foreign direct investment, and amount to more than twice the size of official aid.

The overseas student component of international migration has grown significantly in recent years. There are nearly 5 million students studying abroad worldwide, and they have become an important source of income for many educational establishments. In the UK, international students now make up a quarter of total immigration and contribute more than $35 billion to the UK economy each year.

FIGURE 4.44 West Indian migrants in London in 1950

Source: © Estate of Roger Mayne/Museum of London

Ethnic enclaves and ghettos

International migrants mostly flock to the large towns and cities of the host country. This is because work is more likely to be found in these urban centres and there are probably communities of migrants already established in particular neighbourhoods. These communities are known as enclaves and allow the new arrivals to network. Migrants who move to these areas commonly settle into their new surroundings more quickly as they are surrounded by people speaking the same language and sharing the same culture and religious beliefs.

The US, like Australia, has a particularly long history of international migration and the development of enclaves. During the first half of the 19th century, migrants to Australia were mostly from Britain, Ireland and Germany. They were followed by another wave of European migrants in the late 19th and early 20th centuries, particularly from eastern and southern Europe. At the time, the Italian and East European Jewish settlers became highly segregated from other residents.

Ernest Burgess developed a concentric ring model based on his observations of Chicago in the 1920s to describe urban culture (see figure 4.47). Burgess believed that residential areas grew as a result of invasion and succession. Higher socioeconomic groups moved from the old historic inner city to more desirable outer city commuter areas. Consequently, residential zones would grow by invading and taking over the inner part of the next outer ring. The second ring of his model, being a zone in transition characterised by mixed land use of low-quality housing and new businesses, would accommodate low income families and migrants. Ethnic enclaves were also identified in this model.

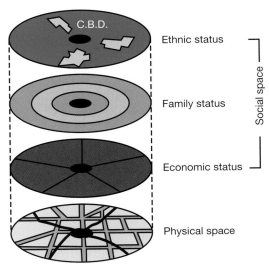

FIGURE 4.45 Murdie's model of residential structure in a city

Source: Robert Murdie, Department of Geography, University of Chicago 1969.

FIGURE 4.46 'Little Italy', New York, circa 1900

> **Ghettos** are defined as extreme forms of residential concentration by culture, religious belief, or ethnicity. To qualify as a ghetto, the area must have a high proportion of a particular group living in that area and the group must also account for most of the population of that area.

The 1930 census data for Chicago (see table 4.11) provides a useful illustration of the difference between a ghetto and an enclave. The Polish migrants were the most concentrated for an individual European ethnic group with 61 per cent of Chicago's Poles living in the Polish district. Furthermore, 54 per cent of the population of the Polish district was Polish. However, the situation was different for the African American population. Ninety-two per cent of the black population lived in the so-called black ghetto, where they accounted for 81 per cent of the population.

FIGURE 4.47 Burgess' model of Chicago

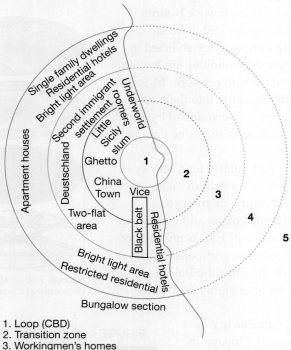

1. Loop (CBD)
2. Transition zone
3. Workingmen's homes
4. Better residences
5. Commuter's zone

TABLE 4.11 Ethnic groups, Chicago, 1930

Ethnic group	Total population of ethnic group	Population of ethnic group in its enclave or ghetto	Total population of the enclave or ghetto	Percentage of ethnic group living in the enclave or ghetto	Ethnic group's percentage of total population of the enclave or ghetto
Irish	169 568	4993	14 595	2.9	33.8
German	377 975	53 821	169 649	14.2	31.7
Swedish	140 013	21 581	88 749	15.3	24.3
Russian	169 736	63 416	149 208	37.4	42.5
Czech	122 089	53 301	169 550	43.7	31.4
Italian	181 161	90 407	195 736	49.7	46.2
Polish	401 306	248 024	457 146	61.0	54.3
African American	233 903	216 846	266 051	92.7	81.5

Source: Thomas Lee Philpott

Ethnic enclaves in the UK

The UK now has a large migrant population (see figure 4.48). Until the 1991 census, a person's ethnic identity was inferred from birthplace and parental birthplace. An ethnic question has since then been included in all census questionnaires. This was specifically designed to monitor equal opportunities and anti-discrimination policies, and assist planning resource allocation and the provision of services.

In 1947, the Indian sub-continent was granted independence from the British Empire. This resulted in the establishment of India and Pakistan as independent dominions but was accompanied by large scale inter-communal violence and dislocation of thousands of people, particularly in the north-east province of Punjab. Many people decided to settle in UK, and the Indian and Pakistani community now accounts for 60 per cent of the Asian population.

Cities such as Leicester in the East Midlands attract large numbers of migrants. By 1951, the ethnic minority population in Leicester numbered about 80 000. In 1972, 20 000 Asian migrants from Uganda also settled in Leicester. The city's minority ethnic population has continued to grow, doubling in size between 1991 and 2011. The largest ethnic minority group is Indian.

The north-eastern quarter of Leicester, which includes the suburbs of Belgrave, Spinney Hill, Stoneygate and Latimer, is now home to a large Asian population (see figure 4.49). These four wards have aggregated minorities that account for over two-thirds of their respective suburb populations. The city's 6500 Polish residents are the third largest overseas-born migrant minority after India and Kenya.

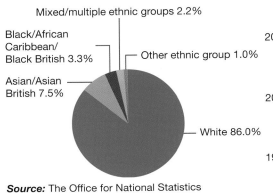

FIGURE 4.48 Ethnic groups in England and Wales, 2011

Source: The Office for National Statistics

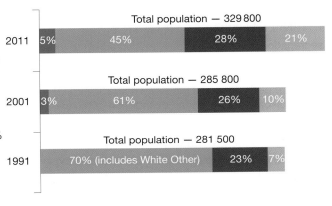

FIGURE 4.49 Proportion of different ethnic groups in Leicester City, 1991–2011

*White Other includes White Irish in 2001 (3600 or 1.3%) and White Irish and White Gypsy or Irish Traveller in 2011 (400 or 0.1%).

Source: The University of Manchester

There are four main factors that have accentuated the demographic structure of ethnic enclaves in the UK. These are:
- net migration
- mortality
- natural increase
- family formation.

Commonwealth migrants to Britain in the 1950s and 60s, described as a replacement population, mostly settled in inner city areas that were already in demographic decline. The inner city was attractive to newly arrived migrants because housing was relatively cheap and there was access to jobs. Later, it was not uncommon for members of the replacement population to follow the path of their predecessors and move into more desirable suburbs.

FIGURE 4.50 Workers' terraced housing in Belgrave, Leicester, UK

The replacement population soon held a larger proportion of the inner city's total population. Migration of the original residents from the inner city areas was age-selective, therefore those who remained were generally older than the younger replacement population. Consequently, the difference in mortality rates between the two populations contributed to the increase in the proportion of the replacement population in the areas's total population. The new migrant population, being younger, also had higher fertility than the population it replaced and this helped to increase its proportion of the total population of inner city areas. People in the replacement population, due to their younger age and cultural traditions, tended to marry early, which meant new family formation for migrants was more rapid compared with that of the original residents. Furthermore, strong pressures to keep the new families close to the parental homes, especially for the Pakistani and Bangladeshi population, further accentuated the degree of ethnic concentration.

The level of residential segregation in European cities between migrants and their hosts is considered moderate and lower than that in the US. Although the perception may be different, there has been a net decrease in segregation as measured by the index of dissimilarity. Upward mobility and new family formation has encouraged the spread and mixing, albeit of different groups. The Indian population in the UK has been notable for its degree of suburbanisation. In London, for example, more than 80 per cent of the Indian community now lives in the outer suburbs.

MEASURING ETHNIC SEGREGATION

Ethnic segregation can be statistically measured by an index of dissimilarity. The index score translates as the percentage of one of two groups included in the calculation that would have to move to other areas in order to produce a distribution that matches that of the larger area. The index score ranges between 0 and 1. When both distributions are perfectly equal the index score is zero. Therefore, the higher the value of the index, the higher the degree of disparity between the distribution of the two populations.

Figure 4.51 shows the growth of population by largest ethnic group in the wards (the local area within a council), where each group is most clustered, and everywhere else between 2001 and 2011. In 2001, the most clustered wards for each ethnic group included a fifth of an ethnic group's total population; the 'less clustered wards' included the remaining four-fifths. The first four columns of the graph relate to the county of Leicestershire, in which Leicester is located, whereas the second four columns only relate to the city.

FIGURE 4.51 Growth of population by largest ethnic group, 2001–11

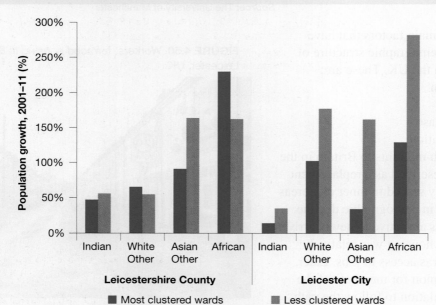

Source: The University of Manchester

Activity 4.5a: Analysing population statistics in the UK

Explain and apply the population data

1. (a) Using data from table 4.12, construct two pie charts to show the proportion of UK residents born in the top ten countries of origin in 2001 and 2011.
   (b) Describe the changes that have taken place since 2001.
   (c) Explain why you think the changes you have identified have taken place.

Analyse the population data

2. (a) Refer to figure 4.51. Is there a difference in growth between those wards where the Indian and Other Asian ethnic groups are most clustered and those wards where they are less clustered?
   (b) Explain whether or not this is evidence of their dispersal towards other suburbs and areas outside the City of Leicester.
3. (a) Describe the spatial pattern of growth between 2001 and 2011 for the African and 'White Other' ethnic groups.
   (b) Why do you think their pattern of growth is different to that of the Indian and Other Asian ethnic groups?

TABLE 4.12 Number and percentage of UK residents born overseas in the top ten countries of origin

Country	2001 census	% non-UK born residents in 2001	2011 census	% non-UK born residents in 2011	% all residents in 2011
India	456 000	9.8	694 148	9.2	1.2
Poland	58 000	1.3	579 121	7.7	1.0
Pakistan	308 000	6.6	482 137	6.4	0.9
Republic of Ireland	473 000	10.2	407 357	5.4	0.7
Germany	244 000	5.2	273 564	3.6	0.5
Bangladesh	153 000	3.3	211 500	2.8	0.4
Nigeria	87 000	1.9	191 183	2.5	0.3
South Africa	132 000	2.8	191 023	2.5	0.3
United States	144 000	3.1	177 185	2.4	0.3
Jamaica	146 000	3.1	160 095	2.1	0.3

Source: The Office for National Statistics

Resources

Video eLesson SkillBuilder: Constructing a pie graph (eles-1632)
Interactivity SkillBuilder: Constructing a pie graph (int-3128)

FIGURE 4.52 Countries in north-west Europe are popular with migrants.

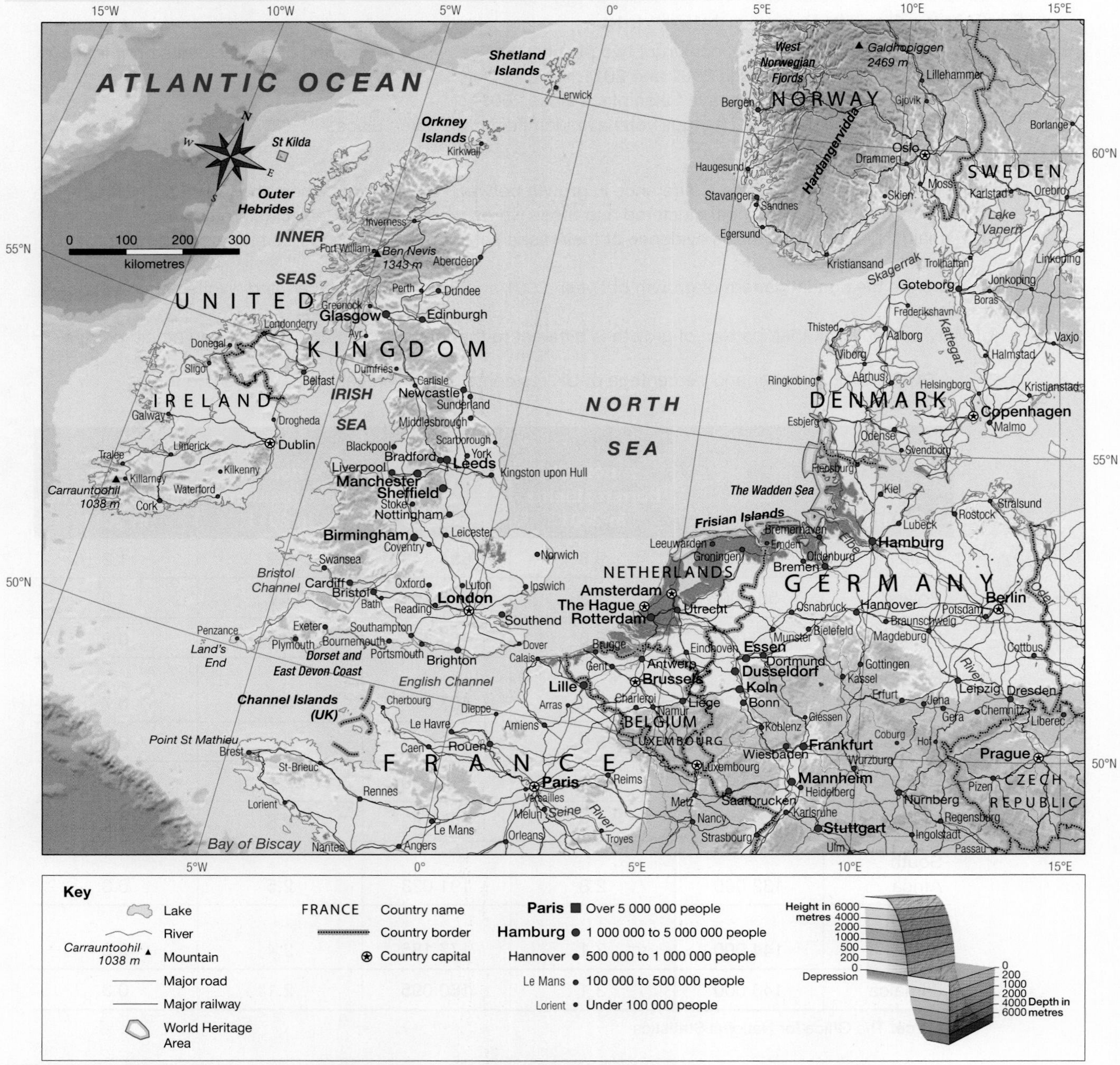

Source: Made with Natural Earth. Free vector and raster map data @ naturalearthdata.com.

Voluntary migration in Europe

In 2015, nearly 20 million of the European Union's birth population lived in a member country in which they were not born. The top three EU countries with the highest number of migrants from other EU countries were Germany, with 5.3 million migrants, the UK, with 2.9 million migrants and France, with 2.3 million migrants.

FIGURE 4.53 EU member states

Activity 4.5b: Analysing the distribution of migrants

Comprehend and synthesise data about migrants

Refer to the Pew Research Centre weblink in the OnResource box.

1. Select the UK and one other host country and compare them in terms of:
   (a) The size of migrant populations from other EU countries
   (b) The geographical pattern of source countries. Do countries in close proximity to the host country seem to attract more migrants than those EU countries further afield?

Explain and analyse your maps

2. (a) Using data from table 4.13 and two blank outline maps of the EU, construct one choropleth map to show the distribution of EU migrants within the EU and another to show unemployment levels. A blank map outline is available in the Resources tab.
   (b) Compare the two patterns.
   (c) Does there appear to be a positive correlation between the number of migrants living in countries and unemployment levels?
   (d) Explain your answer.

TABLE 4.13 Number of residents born in other EU countries and rates of unemployment, 2015

Country	Number of residents born in other EU countries ('000)	Unemployment rate (% working population)
Austria	640	5.7
Belgium	900	8.8
Bulgaria	44	9.7

(continued)

TABLE 4.13 Number of residents born in other EU countries and rates of unemployment, 2015 (continued)

Country	Number of residents born in other EU countries ('000)	Unemployment rate (% working population)
Croatia	70	15.5
Cyprus*	90	15.3
Czech Republic*	160	5.0
Denmark	190	6.3
Estonia*	10	5.7
Finland	140	9.7
France	2330	10.8
Germany	5330	4.5
Greece	330	25.2
Hungary*	300	6.8
Ireland	550	9.5
Italy	1850	11.9
Latvia*	30	9.9
Lithuania*	20	9.6
Luxembourg	220	5.9
Malta*	20	5.1
Netherlands	510	6.8
Poland*	240	7.2
Portugal	190	12.4
Romania	80	6.8
Slovakia*	150	11.1
Slovenia*	70	9.4
Spain	1990	22.2
Sweden	530	7.2
UK	2880	5.5
A10 Countries*		

Source: European Commission, Eurostat

3. With reference to figure 4.54, compare the pattern of total immigration with total emigration
   (a) before 1993
   (b) after 1993.
4. The year 2004 marked the single largest demographic expansion of the EU, with accession of 10 new members. These so-called A10 countries are denoted by * in table 4.13.
   (a) What impact do you think this had on the volume of migration to the UK?
   (b) With reference to figure 4.55, describe the fiscal contribution of migrants to the UK economy.

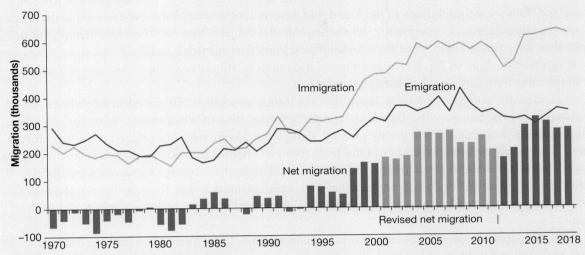

FIGURE 4.54 Migration to and from the UK, 1970–2018

Source: The Office for National Statistics ONS

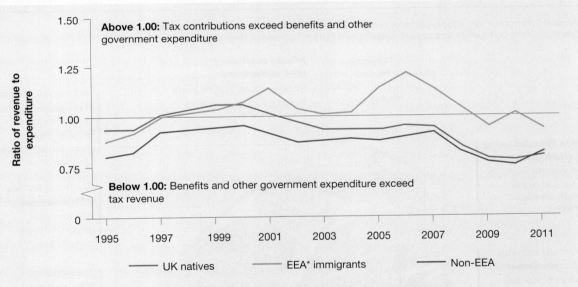

FIGURE 4.55 Fiscal contribution of immigrants to the UK economy

*EEA – the EU plus Norway, Iceland and Liechtenstein

Source: Christian Dustmann and Tommaso Frattini, The Fiscal Effects of Immigration to the UK, The economic journal: the quarterly journal of the Royal Economic Society, 124:580, November 2014.

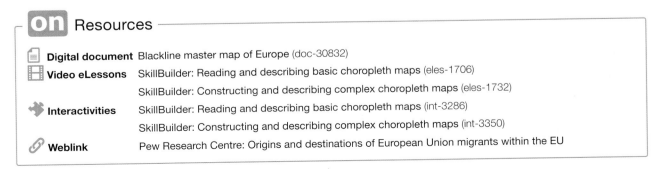

Resources

Digital document Blackline master map of Europe (doc-30832)

Video eLessons SkillBuilder: Reading and describing basic choropleth maps (eles-1706)

SkillBuilder: Constructing and describing complex choropleth maps (eles-1732)

Interactivities SkillBuilder: Reading and describing basic choropleth maps (int-3286)

SkillBuilder: Constructing and describing complex choropleth maps (int-3350)

Weblink Pew Research Centre: Origins and destinations of European Union migrants within the EU

CHAPTER 4 Global population change 233

Migration in Japan

Unlike many European countries, Japan has always been reluctant to open its doors to large-scale immigration. Even though colonial expansion during the first half of the 20th century brought Korean, Chinese and Taiwanese immigrants to the Japan, the government was nevertheless keen to maintain ethnic and cultural homogeneity. Consequently, all former colonial subjects lost their Japanese nationality in 1945. Despite this, a significant number of these immigrants, known as **zainichi**, and their descendants remained. They comprised the bulk of Japan's migrant population until relatively recently as Japan continued to discourage long-term settlement of foreign workers.

Permanent settlement of migrants in Japan remains tightly controlled. This is because children gain citizenship based on their parents' Japanese nationality irrespective of whether they were born in Japan. Naturalisation is available but not readily granted, though the government has relaxed conditions for *zainichi*. This means that the majority of long-term foreign residents in Japan acquire permanent residency and not Japanese nationality. Furthermore, until the early 1980s, these foreign residents were largely excluded from accessing public housing and national health insurance. The foreign share of the overall population has grown slowly and only accounted for 1.8 per cent in 2016, which equates to 2.3 million people.

In 2018, the Japanese government announced legislation that, from April 2019, will see a major relaxation of Japan's tight immigration policy. This could allow as many as half a million foreign workers into Japan by 2025.

FIGURE 4.56 Foreign labour in Subaru's supply chain

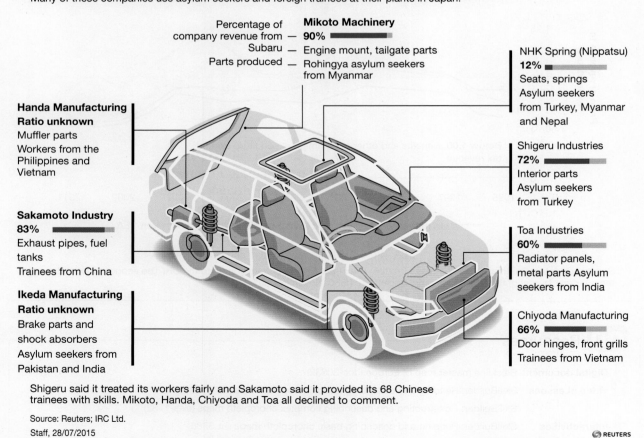

Source: Reuters Graphic / AAP

Japan has a low unemployment rate and some sectors of the economy have struggled to find enough workers to fill the vacancies. The new legislation involves the introduction of a new two tier system based on skill level. Foreign workers possessing certain levels of skill in sectors experiencing labour shortages will be allowed to stay in Japan for up to five years, but will not be allowed to bring their families. Those with a higher level of skill would be allowed to bring their spouses and children. If certain conditions are met, they could be permitted to live in Japan indefinitely. Workers in both categories must pass a Japanese-language exam before admittance.

Japan has admitted foreign workers in the past. Out of a total Japanese workforce of 66 million in 2017, official figures put the number of foreign workers living in Japan at 1.3 million. In reality, the number is a lot less because many of these people are foreign spouses of Japanese citizens or ethnic Koreans long settled in Japan. Therefore, they are not technically workers coming to Japan to seek jobs. Moreover, included in this figure are approximately 300 000 limited-stay foreign students who are allowed to engage in part-time work.

Although immigration to Japan has increased since the 1990s, it is targeted towards those who are deemed beneficial to the country. Refugees remain largely excluded from entering the country. In 2015, fewer than 30 refugees were granted refugee status. However, component suppliers to Japanese car manufacturers, such as Subaru in Tokyo, are allegedly using low paid foreign workers and refugees to meet labour shortages and cut costs (see figure 4.56).

Japan is re-examining its policy on migration because young migrants could provide a solution to some of the economic problems resulting from Japan's ageing population. The IMF has already suggested that ageing is acting as a break on the country's GDP growth. This is because a shrinking and ageing population translates into a smaller domestic market for Japanese products.

According to Japan's National Institute of Population and Social Security Research, the country's population growth had been slowing since the 1980s and eventually peaked in 2008 at 128.08 million. Since then, it has shrunk by more than 1.5 million. In 2017 it dropped by almost 400 000.

4.5.4 Modelling migration patterns

In 1971, Wilbur Zelinsky, an American demographer, devised the Mobility Transition Model (see figure 4.57), which comprises five stages. The first four stages broadly align with the four stages of the demographic transition model and a country's progress along the path to economic development, as witnessed in North America and Europe. The fifth stage was Zelinsky's prediction of what movements would occur in what he called a future super advanced society.

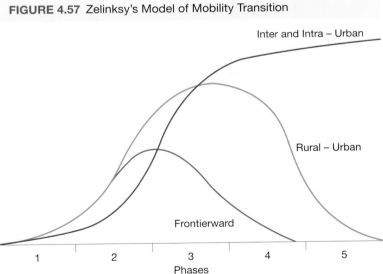

FIGURE 4.57 Zelinksy's Model of Mobility Transition

The stages of the Mobility Transition Model are:
- Stage 1: Pre-modern traditional society aligns with the first stage of the demographic transition. Natural increase is low and mobility is limited to nomadic movements in rural areas.
- Stage 2: Early transitional society is marked by a rising population and surge in rural to urban migration as a result of rapid industrialisation and urbanisation. Migration flows to settlement frontiers and overseas also occur.

- Stage 3: Late transitional society is marked by a slackening rate of movement to settlement frontiers and overseas migration. Rural populations begin to fall and rural to urban migration is replaced by inter-urban and intra-urban migration.
- Stage 4: Advanced society sees the **rate of natural increase** flatten. The volume of migration within urban areas remains high as urban centres merge but counter-urbanisation begins to emerge in response to urban diseconomies and lifestyle-based movements.
- Stage 5: Future super advanced society is marked by mostly inter-urban or intra-urban movement of people as well as a range of short-term migratory movements, which Zelinsky called 'circulation'.

4.5.5 Migration within countries: internal migration

It is estimated that about 3 per cent of the world's population are internal migrants, which is much greater than the world's international migrants. According to the UN, Latin America has a higher rate of internal migration than any other world region.

A major concern for the World Bank is the possibility of up to 140 million people migrating within regions that have experienced ecological disasters such as floods and droughts associated with climate change. For more information watch the documentary about preparing for internal climate migration, go to the **World Bank Groundswell**, available in the Resources tab.

Activity 4.5c: Applying and analysing a correlation in migration intensity

Apply and analyse the data

Demographers at Queensland University's Centre for Population Research have examined migration intensity for a sample of countries over a one- and five-year period. The intensity takes into account all changes of address irrespective of the direction. As demographic mobility is considered an integral part of the development process, a positive correlation would be expected between the two.

1. Using data from table 4.14, use a **correlation technique** (such as a scattergraph and best fit line or Spearman's rank) to examine the relationship between migration intensity and human development.
2. Interpret your result and explain why there is a relationship between the two variables.

TABLE 4.14 Migration versus HDI, 2013

Country	Migration intensity	Human development index
South Africa	17.7	0.67
Sudan	9.9	0.39
Cambodia	13.6	0.56
Indonesia	12.9	0.68
France	26.2	0.88
Ireland	21.2	0.91
El Salvador	16.0	0.67
Peru	19.6	0.73
Mexico	19.3	0.75
USA	23.7	0.91

Source: United Nations Population Division Technical Paper No. 2013/1 - Cross-national comparisons of internal migration: An update on global patterns and trends; United Nations Development Programme Human Development Index 2013, https://creativecommons.org/licenses/by/3.0/igo/

> **Resources**
>
> **Video eLesson** Skillbuilder: Constructing and interpreting a scattergraph (eles-1756)
> **Interactivity** Skillbuilder: Constructing and interpreting a scattergraph (int-3374)
> **Weblink** World Bank Groundswell documentary

Internal migration in Vietnam

Internal migration within Vietnam has increased considerably since the 1980s. This has contributed to the country's mobility transition and increasingly urban population. In 1960, 80 per cent of the population was classified as rural but by 2017 this figure had dropped to 64 per cent.

Vietnamese people used to require permission from local authorities to travel between provinces. Although this is no longer the case, they still require the right *hộ khẩu* to access public services, such as education, health care, vehicle registration, welfare and public sector jobs, outside of their place of origin. The *hộ khẩu* is a residential permit that was used to control where people lived. However, the economic reforms introduced in 1986 aided industrialisation, which in turn increased demand for labour. Unfortunately, *hộ khẩu* registration is administratively cumbersome so many migrants residing in the main urban centres, such as Ho Chi Minh City, Hanoi, Da Nang, Can Tho and Hai Phong, don't have a *hộ khẩu* or do not bother updating it to reflect their new permanent place of residence. Part of the problem stems from the fact that many of those without a *hộ khẩu* who reside in the main urban centres are originally from rural areas, where there is a lack of understanding of government processes. There is also the problem of the lack of coordination between provincial authorities when it comes to public records. The Vietnamese government is currently pursuing an annual economic growth target of 6 per cent until 2035 and is planning to replace the *hộ khẩu* system with a new national citizen database in 2020. This would free up a lot of labour currently locked into its agricultural sector.

Between 1999 and 2009, the internal movement of people in Vietnam doubled from 4.3 million to more than 8 million (see figure 4.58). Since then, the volume of migration has continued to grow and the direction of movement change as shown in table 4.15. Regionally, south-east Vietnam has the greatest proportion of the migrants, with Ho Chi Minh City recording the highest intake. International migration has remained low, involving less than 3 per cent of the population.

FIGURE 4.58 Net internal migration rate, Vietnam, 2004–9 (%)

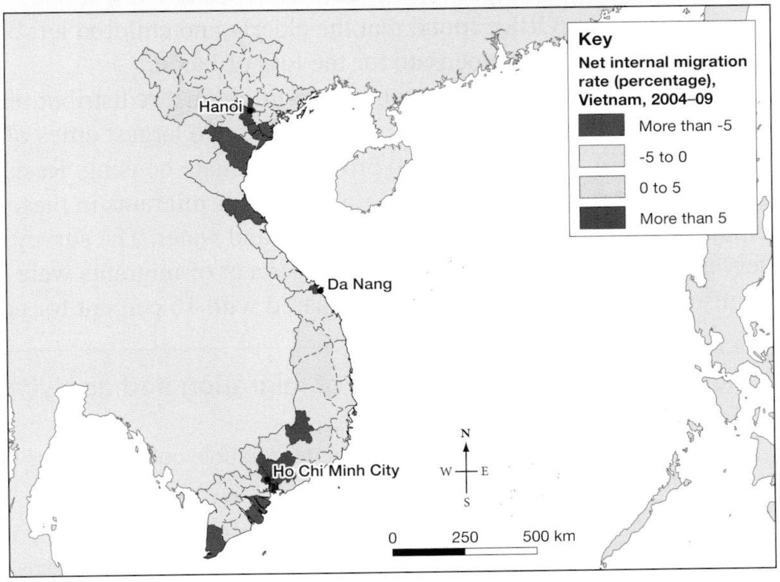

Source: Made with Natural Earth. Free vector and raster map data @ naturalearthdata.com.

TABLE 4.15 Types of migration in Vietnam, 1999–2015 (%)

Period of census (years)	Moves within rural areas (rural–rural)	Moves from rural to urban (rural–urban)	Moves within urban areas (intra–urban)	Moves from urban to rural area (urban–rural)
1999–2009	33.7	31.6	26.3	8.4
2010–2015	19.6	36.2	31.6	12.6

Source: United Nations Population Fund, 2008

As in other countries, Vietnamese migrants tend to be in younger age brackets. The average age of migrants in Vietnam is 29 years, with most in the 15–39 years range. Although men represent just under half of all migrants aged 15–59 years, they are more likely to migrate for work purposes than women, who are more likely to move for family or education. The Kinh and Hoa people, from the north and south of the country respectively, migrate more than any other of the 52 ethnic groups. Kinh Vietnamese make up 86 per cent of the country's population whereas the Hoa people are Chinese-Vietnamese and only constitute 1 per cent of the population. However, the Hoa people are highly culturally assimilated with the ethnic Kinh group. Prior to the end of the Vietnam War in 1975 and the establishment of the communist government, these groups formed a very prosperous middle class in Vietnamese society.

Migration has an impact on economy. This is because most migrants send money back home to their families, who spend it within their country. According to the World Bank, almost all migrant-sending households consider migration to have had a positive impact on household income. However, in 2016, the General Statistics Office found that the elderly and children left behind generally have to undertake more agricultural work to compensate for the loss of labour.

Migration flows within Vietnam are resulting in a redistribution of the country's population to regions that are the location of industry, including the two largest cities of Hanoi and Ho Chi Minh City. This has tested the ability of such cities to provide adequate housing. Results of the 2015 National Internal Migration Survey found major points of dissatisfaction for migrants in these cities were the high rent needed for accommodation, and the cost of electricity and water. The survey also found that migrants were living in very cramped conditions. More than 40 percent of migrants were housed in accommodation with less than 10 m^2 of living space per person, compared with 16 percent for non-migrants.

> **Activity 4.5d: Analysing types of migration and applying understanding**
>
> **Explain and analyse the data**
>
> 1. Using data from table 4.15, construct four double column graphs to show how the direction of migratory movement has changed since 1999.
>
> **Apply and communicate your knowledge of types of migration**
>
> 2. To what extent do you think Vietnam appears to conform to Zelinsky's model of mobility transition?

Internal migration in the UK

The volume of migration in the UK is broadly similar to that of Vietnam. Between 2009 and 2015, 8 million people changed address, making the UK's annual **migration rate** of 3.5 per cent one of the highest in Europe. However, the majority of moves were of short distance to neighbouring districts.

Some of the UK's largest cities are experiencing a net loss in migration. Between 2009 and 2013, Birmingham lost almost 14 000 people, while both Manchester and Leicester lost more than 8000. Liverpool and Sheffield experienced net migration losses of more than 4000 each. This trend may be surprising because cities have higher jobs growth and better wages, although they can suffer from lower housing affordability.

Since 1991, the population of London has been steadily growing, reversing a downward trend that started in 1939 (see figure 4.59). In recent years, population growth has been particularly rapid. From 2011 to 2016, the population of London grew by 7.5 per cent, twice as fast as the UK as a whole, and reached a population of 8.8 million. The population of London is projected to continue growing over the coming decades and is expected to increase by more than half a million between 2016 and 2021, not as a result of net migration, which has been negative for some years, but instead by natural increase. London attracts the young but loses the older age groups. As in Vietnam, the major pull factors of cities are higher incomes and greater employment opportunities, although they can be offset by higher rents and general living costs.

The regions closest to London, namely the South East and East of England, have experienced a net internal inflow of migrants — more than 160 000 in 2014 — with the South West region close behind. These regions have higher populations in general than others and greater net migration as a proportion of

FIGURE 4.59 London population growth in millions, 1931–2021

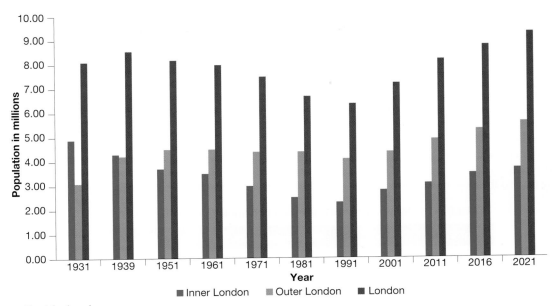

Source: Trust for London

FIGURE 4.60 Regions in the UK

their respective populations. The UK's census body, the Office of National Statistics (ONS), credits this to the higher quality of life in the south and the proximity of London.

The ONS also found a positive correlation between the inflow and outflow of people. This is because areas with more migration have larger populations to start with. Cities will generally have the highest turnover of people given their dense populations.

Activity 4.5e: Describing and comparing patterns of movement
Explain and analyse the data

1. (a) Using data from table 4.16, construct a choropleth map to show net moves per thousand in the UK.
   (b) Describe the pattern of movement.
   (c) Compare the pattern of movement in 2015 with that of 2012 in figure 4.61.
2. (a) Refer to figure 4.62 and describe the pattern of migration flows from London to other regions in England and Wales.
   (b) Why do you think regions close to London received more migrants than more distant regions?

TABLE 4.16 Regional net migration and unemployment in the UK, 2015

Region	Migration net moves ('000)	Unemployment
South West	+4.8	1.4
East	+2.4	1.6
South East	+1.8	1.3
North East	−0.1	3.8
East Midlands	+2.0	2.1
West Midlands	−0.4	2.7
North West	−0.5	3.8
Yorkshire and Humberside	−0.4	3.0
London	−9.1	1.9
Scotland	−0.2	2.9
Wales	+0.1	3.1
Northern Ireland	0.0	4.5

Source: The Office for National Statistics ONS

FIGURE 4.61 Internal migration by local authorities in England and Wales, to June 2012

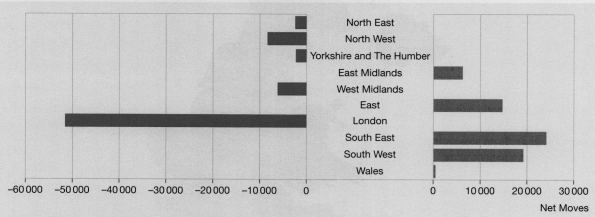

Source: The Office for National Statistics ONS

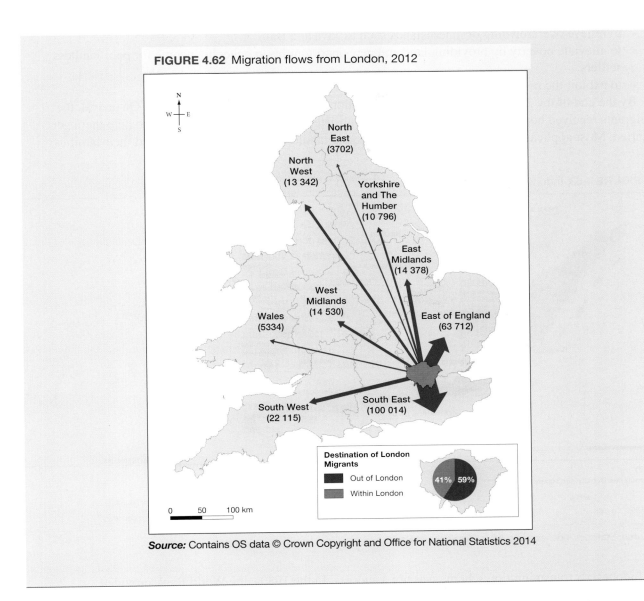

FIGURE 4.62 Migration flows from London, 2012

Source: Contains OS data © Crown Copyright and Office for National Statistics 2014

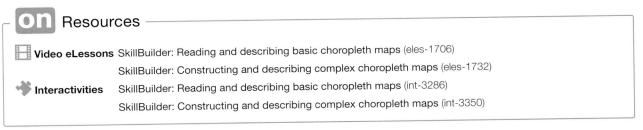

Planned migration

Planned migration is usually the coordinated movement of people to better use resources or fill labour shortages. This is overseen by a government body. In 1905, during colonial rule of what is now Indonesia, the Dutch began an experimental program of colonising and developing the outer islands with peasant farmers from the densely populated island of Java (see figure 4.63). This process is known as **transmigration**, and was taken over and expanded by the newly formed Indonesian government after independence in 1949. The Indonesian government's goals were:

CHAPTER 4 Global population change 241

- to relieve overcrowding of inner islands such as Java and Bali
- to alleviate poverty by providing land and new opportunities to generate income for poor landless settlers
- to exploit the resource potential of the outer islands.

By the end of the 20th century, more than 3.6 million Indonesians had been resettled. On arrival, the migrants received houses, land for farming, and a subsistence and production package to help them get started. Most applicants for transmigration were young landless agricultural workers and their families.

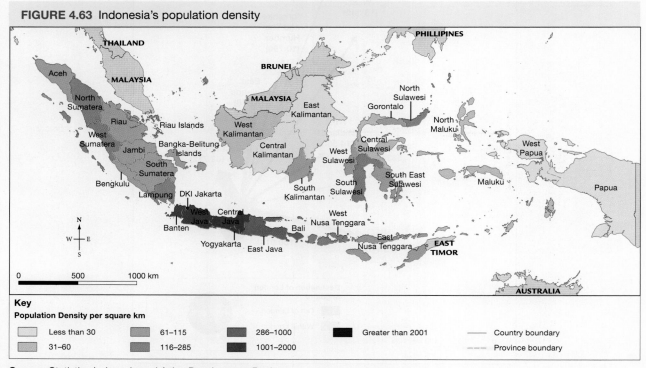

FIGURE 4.63 Indonesia's population density

Source: Statistics Indonesia and Asian Development Bank

Indonesia's transmigration scheme was not a success. In the 1980s, the World Bank provided considerable financial support to help increase the number of migrants leaving Java. However, in the following decade, there were environmental, social and economic concerns, and the scheme failed to achieve targets, and financial assistance to support further resettlement began to dry up. The criticisms of Indonesia's transmigration scheme can be summarised as follows.

- Its impact on Java's population density was minimal.
- Poor farming conditions and lack of infrastructure did not alleviate poverty, only redistributed it.
- Land clearing of the outer islands threatened 10 per cent of the world's remaining rainforest.
- Resettlement was expensive and increased the country's foreign debt.
- The scheme camouflaged the government's desire to politically control the outer islands.
- The scheme violated local land rights.
- Wetland drainage increased carbon emissions from dehydrated peat soils.
- Demise of traditional ethnic culture in the outer islands.

It would appear that some of the coercive elements of transmigration have waned. However, the government's desire to exploit natural resources to generate much needed revenue may trigger a new wave of migration.

West Papua is an ideal region for continued transmigration efforts in Indonesia. West Papua comprises 24 per cent of Indonesia's total landmass but only 1.7 per cent of the country's population. It is well-endowed with natural resources, including the largest extant tracts of rainforest in south-east Asia, vast oil and gas reserves, and large deposits of copper and gold. In 1959, only 2 per cent of the population was

FIGURE 4.64 Transmigrants in West Papua

from outside the province, but by 2011 the proportion of migrants had reached 53 per cent (see figure 4.64). Critics argue that the continuing transmigration program has led to large numbers of migrants controlling the economy in towns and villages. Consequently, local Papuans feel increasingly marginalised, which has led to violence. In early 2019, direct conflict in the mountainous Nduga region led to the violent deaths of an estimated 20 migrant road construction workers. This sparked retaliation from the Indonesian Army, which is endeavouring to suppress an independence movement in the province.

Forced internal migration in Cambodia

Since the 1970s, the Cambodian people have experienced widespread forced migrations. During the early 1970s, there was large scale migration to Phnom Penh, the capital city, from rural areas to escape being victims of the Vietnam War. However, when the Marxist regime of Khmer Rouge seized control of the country in 1975 its leader, Pol Pot, set about creating a communist 'agrarian utopia'. He admired the self-sufficient tribal system of Cambodia's rural north-east, which he likened to communes. He immediately embarked on a program to replicate this and renamed Cambodia 'Kampuchea'.

It is estimated that over a third of the country's population was forced to relocate from urban to rural areas during the Khmer Rouge regime. Twenty thousand people died in the forced evacuation of Phnom Penh, through starvation, disease, executions and beatings. On arriving in the country, migrants were forced to create new collective farms. The death toll from almost four years of Khmer Rouge rule has been estimated at between one and three million.

The invasion of Cambodia by Vietnamese troops led to the overthrow of the Khmer Rouge in 1979 and the establishment of a more moderate communist government. Phnom Penh was repopulated in the 1980s, primarily by rural migrants moving into houses and onto land but in an unregulated manner. Occupation rights were gradually recognised but the government did not recognise property rights until 1989, which later became enshrined in the country's constitution. However, this did not prevent 30 000 residents being forcibly evicted from the centre of the Phnom Penh between 1998 and 2003 by corrupt officials.

Growing competition for land and natural resources in rural areas has led to an increasing number of forced evictions. In 2012, thousands of people in Koh Kong province were relocated to new villages in the mountains to make way for a Chinese resort development. The $US 3.8 billion tourism project stretches across 45 000 hectares of pristine coast along the Gulf of Thailand, 400 km west of Ho Chi Minh City.

Planned Israeli settlement of the West Bank

Israel (see figure 4.65) has occupied the west bank of the Jordan River in Palestine since the Six Day War between Israel, Egypt, Syria and Jordan, in 1967. Since then, more than 600 000 Israeli settlers have moved into the West Bank territory, which is also home to 3 million Palestinians. Israelis are housed in defensive fenced townships or colonies of varying sizes. Israel's Supreme Planning Council in the Civil Administration is responsible for the planning and organisation of its settlements in the West Bank. Israeli governments have been reluctant to return the territory to Palestine, despite Article 49 of the Geneva Convention, which forbids a country from deporting or transferring part of its own population to the territory it has occupied.

The number of Israeli settlers migrating to the West Bank looks set to increase. In late 2018, the construction of another 1451 housing units was formally approved by the Civil Administration, as were plans for the construction of 837 additional units, some of which will be in relatively isolated settlements. With 11 per cent of Israel's population now living in the West Bank, further migration of settlers could make a long-term peace agreement in the region increasingly difficult to achieve.

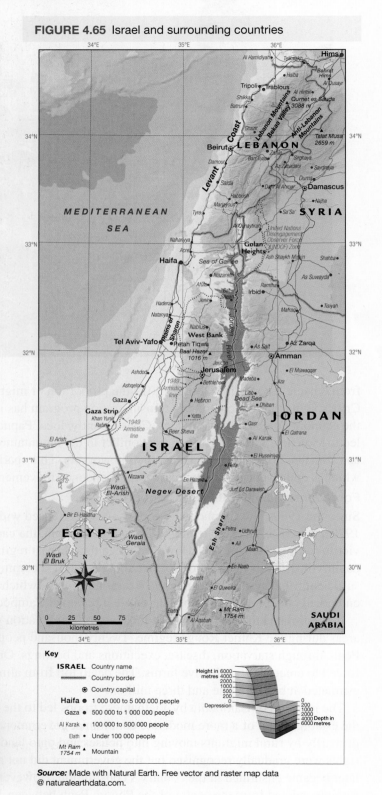

FIGURE 4.65 Israel and surrounding countries

Source: Made with Natural Earth. Free vector and raster map data @ naturalearthdata.com.

FIGURE 4.66 Construction continues in the Israeli settlement of Gilo.

4.6 Review

4.6.1 Chapter summary

This topic has covered some key points about global population change.

Global patterns of population growth

- Census data enables governments to make relatively accurate future population forecasts and plan for population growth or, in some cases, population decline.
- The UN has estimated that the world population is likely to grow to nearly 10 billion by 2050.
- The world fertility rate — the number of children born per woman of childbearing age — has declined from over 5.0 in the mid-1960s to just under 2.5 now.
- Fertility rates vary between world regions and between countries. Least developed countries will continue to have relatively high levels of fertility but they are declining.
- Declining mortality (death rates) contributes to population growth. Improvements in health have been significant factors in recent years to declining mortality.
- The reduction in global fertility and mortality is not only slowing the pace of population growth but is also producing an older population.
- The age–sex structure of a population is important because it defines the relative numbers of young and old, and the balance of males and females, which in turn influence the overall number of births and deaths.
- Rostow's Five Stages of Economic Growth suggested that after initial capital investment, countries would embark on an evolutionary process lasting about 60 years in which they would move up through five stages of a development.
- In 1974, Immanuel Wallerstein's *The Modern World System* was published. This is a theoretical framework comprising four stages in which he attempted to explain how the modern capitalist world economy evolved from the age of feudalism to the present day.
- In the early 20th century American demographer Warren Thompson devised the demographic transition model, in which countries transitioned from having high birth and death rates to low birth and death rates as they became increasingly industrialised and democratic.
- The four-stage transition model which builds on Thompson's model has been a useful tool for analysing patterns of population dynamics.

An ageing world

- Challenges of an ageing population include the dependency ratio.
- The dependency ratio is the number of dependents in a population divided by the number of working age people. The higher the ratio, the greater the economic burden carried by working age people.
- After three decades of the One Child Policy, China now faces the problems of a shrinking workforce and an increasingly ageing population.
- In 2018, the number of births recorded in Japan fell to its lowest since records began in 1899.

Patterns of changing population distribution and density

- Physical and socioeconomic factors have influenced both the distribution and density of the world's population.
- In 1950, about two-thirds of the world's population lived in developing countries. By 2030 this figure is expected to have increased to 85 per cent.
- Although Asia is by far the most populous continent, Africa is now the fastest growing region in the world and its share of global population is expected to rise to 20 per cent by 2050.
- Demographic shifts in population distribution will impact different parts of the world in different ways.

People on the move: international and internal migration

- Global migration is increasing as a result of increasing connectivity, continued regional inequalities and demographic imbalances.
- Migration can be forced or voluntary.

- Reasons for forced migration include natural hazards and disasters, dam construction, conflict, persecution (religious or cultural), and food and water insecurity.
- Voluntary migration is when a migrant has a choice as to whether or not they wish to migrate to a particular place. Reasons for voluntary migration can be economic, education, social, environmental, or related to governance.
- Migration can be internal (within a country) or international (to a different country). Rates of internal migration are higher than international migration.
- Migration has social, economic and demographic impacts on the host area and the home area.

4.6.2 Key questions revisited

You should now be able to answer the following questions.
- Why is accurate and reliable census data useful to governments?
- What trends and patterns can be identified in global population change?
- Which demographic concepts are critical in explaining population change?
- How can models help us understand the past and possible future changes in population across the world?
- What spatial patterns of population distribution and density result from population change?
- How do countries manage the challenges of changing characteristics of their populations such as ageing?
- What are the positive and negative impacts of migration of people on places of origin and destination?

4.6.3 Practice Assessment 4

Go to your Resources panel for a practice assessment for this topic, along with a stimulus sheet and marking guide.

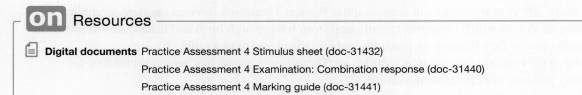

Resources

📄 **Digital documents** Practice Assessment 4 Stimulus sheet (doc-31432)

Practice Assessment 4 Examination: Combination response (doc-31440)

Practice Assessment 4 Marking guide (doc-31441)

GLOSSARY

activism efforts to affect societal change through grass roots action

age of high mass consumption characterised by a highly urbanised consumer economy

age–sex pyramids see **population pyramids**

age–sex structure the composition of a population by age (0–4 years, 5–9 years, 10–14 years ... 80–84 years, 85–89 years, 90+ years) and sex (male, female)

albedo a measure of the ability of surfaces to reflect sunlight (heat from the sun). Light surfaces return much of the heat back to the atmosphere and dark surfaces absorb heat from the sun.

algal blooms rapid increases in the accumulation of algae in a water body due to some external pressure (e.g. rapid increase or decrease in temperature or increased nutrients)

amenity the desirable or useful features or facilities of places including access to shops and other services required for daily living, employment, health care, educational services, transport, cultural and leisure services, and green spaces

angiosperms plants that produce flowers and develop seeds within a contained ovary. Most grasses, shrubs and trees are angiosperms

anthropogenic biome the human biome, created by human interaction with the biosphere

aquaculture aquatic animals such as fish, shellfish, plants, algae, and other creatures are bred, raised and harvested for food or some other purpose

archipelago an island group or chain of islands, such as Indonesia

asylum seeker a person who has fled a country involuntarily and sought protection as a refugee, but whose claim for refugee status has not yet been assessed

baby boom an upward spike for a number of years in birth rates; the period in Australia from the late 1940s to the early 1960s when birth rates were higher than they had been since the 1920s

biodiversity biological diversity; describes the variation of living plant and animal species that occupy an area or ecosystem

biogeographic areas smaller regions where the distribution of plants, animals and geology reveals many shared or common features, such as Wallum country of southern Queensland

biomes very large regions of the Earth where area-specific plants and animals have adapted to the climate, soil and relief of that environment. Deserts and rainforests are biomes.

birth rate also known as crude birth rate; the annual number of live births in a particular place per 1000 people

broadscale clearing indiscriminate clearing of large tracts of land, usually to prepare that land for agricultural production. A heavy chain strung between two heavy bulldozers is the favoured method of broadscale clearing in Queensland.

brown–field developments re-development of sites within urban areas previously used for another purpose, often industry or warehousing

carbon cycle the chain of biogeochemical processes that exchange or move carbon between the biosphere, the atmosphere, hydrosphere and lithosphere

carbon sequestration the process of carbon being removed from the atmosphere and stored as either a solid or a liquid

chaparrals the hot, dry shrublands and heath country in the US and South Africa

climate change changes in the physical systems (biosphere, hydrosphere, atmosphere, lithosphere) that change the Earth's climate; the term used to describe climate patterns that have become significantly different from what regularly occurred during past centuries. It is often attributed to increased amounts of water vapour, CO_2 and methane in the atmosphere due to anthropogenic factors.

continents the largest landmasses on Earth: Africa, Antarctica, Asia, Australia, Europe, North America and South America

core describes the advanced economic core of a country or region surrounded by the less economically developed periphery

Coriolis effect a force like inertia that deflects objects moving across the surface (air, ocean currents, etc) to the left in the southern hemisphere and to the right in the northern hemisphere. It is caused by the rotation of the Earth.

crops plant or animal products that can be grown and harvested for consumption and/or profit

cumulative causation see **multiplier effect**

curvilinear relationships relationships in a correlation that are more complex than simple linear relationships. In the higher ends of each variable in a curvilinear relationship there is little variation whereas at the lower end there is more variability.

death rate also known as crude death rate; the annual number of deaths in a particular place per 1000 people

deforestation the intentional clearing or removal of forests to make way for some other purpose such as farming, housing, constructing a dam, and so on. Deforestation is regarded as a permanent loss of forest and is most common in countries with large areas of rainforest such as Brazil and Indonesia.

demographic transition describes how changes in population growth respond to relative differences in birth and death rates, which themselves change over time

demographics the statistical study of populations or groups of people

demography the study of population, especially human population. Demography involves statistical analysis of characteristics such as population size and composition, distribution across space and the processes through which populations change over time.

dependency ratio the percentage of a population dependent on people who are in the workforce. Statistically, the dependency ratio is the percentage of a population aged below 15 years and above 65 years.

developed world countries with mature industrialised economies, which exhibit a high gross domestic product (GDP) per capita.

development the process of changing the purpose of a piece of land by constructing or removing something

diversity a variety of different things or species

downwelling the downward movement of ocean water from the surface to the ocean floor. It happens when surface water becomes very cold and saltier than normal. This makes it denser and it drops downwards. Downwelling occurs mostly when a current reaches a continent or near the poles.

drive to maturity increasing standards of living and a diversifying economy lead to social changes, such as later marriage, which bring about a slackening of population growth, a direct consequence of a declining birth rate

drought escapers plants that have developed specific reproductive and osmotic features that allow them to only grow and flower for short periods after rain, such as ephemerals. An example is Sturt's desert pea, the floral emblem of South Australia.

drought resisters plants with root and leaf features that ensure survival during prolonged periods of dry weather. They include cacti (thin needle-like leaves) and pineapple.

ecological climax when a community or ecosystem experiencing change (e.g. after a fire) has redeveloped over time to where its plants, animals and abiotic features are in a state of balance (equilibrium).

ecosystems communities of biotic (living organisms) and abiotic (non-living things like water and soil) that are linked through nutrient cycles and energy flows. An ecosystem may be large like a barrier reef or small like a garden.

evapotranspiration the process of water transferring from the land to the atmosphere via evaporation and transpiration

fertility rate the average number of children per woman of childbearing age (15–49 years) during her lifetime

Five–Year Plan (China) a series of strategic social and economic development initiatives based on a five-year timeframe and goal. China is currently using its 13th Five-Year Plan.

Food and Agricultural Organization (FAO) a specific agency of the United Nations that leads international efforts to reduce hunger, particularly in developing countries, by using expertise and skills from developed countries.

food insecurity a situation in which people do not have physical and economic access to the basic food they need to work and function normally

forced migration when people are forced to leave an area against their will or as a result of life-threatening circumstances

fossil fuels fuels formed by natural processes that contain energy originally captured during photosynthesis (e.g. decomposition of living matter into coal or oil)

gentrification a process of demographic and aesthetic change in urban areas where an area becomes desirable, wealthier residents move in and poorer residents move away due to increased property prices

geospatial technologies tools that allow the user to collect, manipulate, distribute, analyse and communicate geospatial information such as GPS receivers, GIS applications or UAV drones

ghettos extreme forms of residential concentration by culture, religious belief or ethnicity

global warming the gradual increase in temperature in the Earth's atmosphere. Over the past half century, atmospheric temperatures have increased at the fastest rate in recorded history, generally attributed to increased levels of CO_2 and other greenhouse gases.

GNI per capita ($US PPP) a measure of a country's relative wealth in current international dollars — gross national income (GNI) in purchasing power parity dollars ($US PPP) divided by the mid-year population (per capita)

greenfield developments urban development in locations that were previously non-urban, either natural or used for primary industry

greenhouse gases gases such as water vapor, carbon dioxide, methane, nitrous oxide and ozone found in the atmosphere. Because they absorb radiant energy and trap heat, they help warm the Earth and cause the greenhouse effect.

gyres large rotating ocean currents. There are five major gyres in the world: the North Pacific, North Atlantic, South Pacific, South Atlantic and Indian Ocean gyres.

Hadley Cell the vertical circulation of warm tropical air high into the troposphere, usually driven by trade winds from subtropical areas

heat sink a sink that absorbs heat

Holocene period the period from the end of the Pleistocene (approximately 11 000 years ago) to today

Humboldt Current a cold, low-salinity ocean current that flows north along the western coast of South America towards the equator

hydrological cycle the process of exchanging water between the air, land and sea through evaporation, condensation and precipitation

import substitution policies replacing imports with domestically-produced goods

Industrial Revolution the period of invention that started in England during the late 1700s and changed the nature of manufacturing from a domestic system to a factory system

inorganic compound a chemical compound that does not involve a carbon-hydogen bond, such as water and carbon dioxide

inter–state migration the movement of people from one state to another state

Intergovernmental Panel on Climate Change (IPCC) the leading international body and source of scientific data and technical information used by the United Nations for assessing climate change. It was formed in 1988.

Intertropical Convergence Zone (ITCZ) the large region of low pressure close to the equator where the trade winds of both the northern and southern hemispheres meet. Due to high temperatures, warm ocean water and trade wind convergence, the region has unstable air that rises, forming thunderstorms. It moves either side of the equator according to the summers and often appears on weather maps as a monsoon trough.

intra–state migration the movement of people from one place to another within a state, such as from a rural area to a town or city

inundation describes when water rises and floods an area that is not normally below water. Land may be inundated after heavy rain.

land cover literally what is covering the land (e.g. vegetation, human infrastructure or development, agriculture or bare earth)

least developed countries the world's poorest countries; the countries with the lowest GNI per capita and relatively lower life expectancy and higher birth, death and fertility rates, mostly located in Africa

leeward side the downwind side not facing the direction from which a wind is blowing. The leeward side of a mountain range is usually drier than the windward side.

life expectancy the average number of years a newborn infant is expected to live, given the mortality rates at the time of its birth

littoral drift the movement of sedimentary materials into the littoral zone (shoreline) under the influence of waves and tides

location quotient a ratio used to indicate the concentration of a demographic feature in a particular location, such as a suburb compared with a larger benchmark region — a city, a state or the country as a whole.

migration rate immigration (incoming) number minus emigration (departing) number per 1000 people

Millennium Development Goals the eight goals devised by UN member nations in 2000 to help countries fight poverty, hunger, disease, illiteracy, environmental degradation, and discrimination against women. From 2015, these were superseded by the UN's Sustainable Development Goals.

minerals naturally occurring solid chemical compounds (e.g. salt, quartz)

Modernisation Theory developed in the mid-20th century with the aim to show that low income countries needed to follow the same path to development as the West

monsoon describes the seasonal changes in atmospheric circulation and precipitation in warm regions. In northern Australia, the monsoon season lasts from December to March and is often associated with an inflow of moist north-westerly winds that generate heavy rainfall and cyclones.

Montreal Protocol the agreement signed by 24 countries in 1987 in London to reduce production and emission of greenhouse gases

more developed countries the world's wealthiest countries; countries with relatively high GNI per capita and demographic features such as high life expectancy and low birth, death and fertility rates

mortality rate death rate, usually for a specific group in a population, for example, infant mortality rate, child mortality rate, maternal mortality rate

multiplier effect the snowballing of positive benefits associated with the establishment of a new economic activity in a place. The newly established activity attracts associated economic activity which in turn attracts further activity.

national parks areas of land set aside by the government to protect its biodiversity, plants, animals and landforms, such as Springbrook National Park, near the Gold Coast

net migration the number of immigrants minus the number of emigrants

newly industrialised countries (NICs) countries where the economy has transitioned from that of a less developed country based on primary products to one increasingly based on the production of manufactured goods. They are not as highly developed as countries with mature economies.

nitrogen cycle the biogeochemical loops where nitrogen in its various forms is circulated between the atmosphere, hydrosphere, lithosphere and biosphere

nomadic herding a form of animal grazing where herders continuously move their cattle, goats, horses, sheep or camels to fresh pastures. Herders do not own the land but they have the right to use it.

nutrients substances used by an organism to survive, grow and reproduce

oceans the largest bodies of water on the Earth, holding more than 96 per cent of all water. Ocean water is approximately 3.5 per cent salt.

percentage urban the percentage of the population living in areas termed urban (towns and cities) by a country or by the United Nations Organization

permafrost any ground (rock, soil, sediment) that is frozen or remains 0 °C or colder for a period of two years. It is most common in regions in the high latitudes, like the tundra. Global warming is threatening to thaw out large areas of permafrost.

photosynthesis the process of plant growth where sunlight triggers the conversion of water and CO_2 to carbohydrates

phytoplankton microscopic marine organisms, such as diatoms, that live close to the sea surface where photosynthesis occurs. Krill feed on phytoplankton.

planned migration the coordinated movement of people to better use resources or fill labour shortages, overseen by a government body

Pleistocene Ice Age the last great ice age, lasting from more than 2 million years ago until around 11 000 years ago. During this time, large parts of the northern continents were covered by extensive ice sheets and glaciers filled many valleys. Sea levels were much lower than at present.

polar vortex a form of upper-level low-pressure region swirling about near the Earth's poles. In winter, the polar vortex at the North Pole increases, sending icy cold air to populated regions in Canada and the US.

polders low-lying areas of land often surrounded by drains, canals and walls to keep back the sea and control flooding

population composition the various elements that make up a population, including males and females, age groups, Indigenous peoples, Australian and non-Australian born people

population density a measurement of the number of people located in a given area, usually the number of people per square kilometre

population distribution the way a population is spread over space (sparsely populated, densely populated, etc)

population momentum growth the continued growth of a population despite the fertility rate being below replacement level. It ceases once the segment of the population of reproductive years has moved beyond childbearing age.

population pyramids also known as age–sex pyramids; a compound bar graph illustrating the age–sex structure of a place's population

population size the total number of people located in a particular area

prairies vast stretches of temperate grassland with few trees. They are fertile due to the thousands of years of decaying grasses. Because soils are highly compacted with few air spaces, it is difficult for trees to grow.

precipitation rainfall

preconditions for take-off in which western development aid and investment would assist improvements in agriculture and economic infrastructure, such as roads, and therefore encourage further overseas investment by foreign companies

primary data data collected by the author directly from the source

pull factors the positive features of places that are attractive to migrants; they draw people towards places

push factors negative features of places that are responsible for people leaving

rain shadow a dry zone on the leeward side of a mountain range. Because the mountains often block any moist on-shore flow of wind, the rain shadow remains dry.

rangelands a broad term describing remote country used for grazing domestic livestock or wild animals. They include tallgrass and shortgrass prairies, semi-desert grasslands, shrublands, woodlands, savannas, chaparrals and steppes. Much of Australia's inland could be classed as rangeland.

rate of natural increase the birth rate minus the death rate, expressed as a percentage: (crude birth rate minus crude death rate) × 100/1000

remote sensing the method of using distance technology (satellites, cameras, aircraft) to gain information about a place or item from a distance

replacement fertility the rate required for a population to replace itself from one generation to the next (approximately 2.1 births per woman)

resources sources from which some benefit is produced (e.g. wood, minerals, water)

riparian zone the interface between a waterway and the land that provides habitat, prevents erosion and blocks nutrients and sediments from entering the waterway

salt intrusion where saline water moves into freshwater aquifers resulting in contamination of drinking or irrigation water sources. Storm surges during cyclones or tsunamis can carry ocean water into fresh groundwater areas.

savanna the region of tropical grasslands intermixed with woodlands. Trees are widely spaced and there is no distinct canopy.

secondary data data collected by someone other than the author

sequestration the removal and long-term storage of carbon dioxide and other carbon materials to help mitigate the effects of climate change

sink either a natural or artificial reservoir that collects and stores carbon materials. Forests and oceans are natural sinks.

socioeconomic status a measure of social status based on economic and social data such as income, occupation and education level

soluble gas a gas that can dissolve in water. The degree of solubility varies according to air or water temperature.

Southern Oscillation Index (SOI) a measure of the development and intensity of El Niño or La Niña events in the Pacific Ocean. It is based on the pressure differences between Tahiti and Darwin.

steppes flat grassland with no trees

sustainability a process of maintaining balance in an environment while retaining the ability of current and future inhabitants to use that environment

synoptic chart a weather chart that shows a summary of weather detail over a large area. For example, the daily weather map of Australia shows a summary of air pressure, rainfall, winds, temperature, and so on.

take-off a relatively short industrialisation period of rapid economic growth in which a new urbanised entrepreneurial and middle class of consumers emerges that generates more wealth as a result of cumulative causation

thermocline the cool layer of ocean water below the surface. It is still warmer than the very cold layer on the ocean floor.

thermohaline describes water circulation that is like a giant water conveyer belt that returns wind driven ocean surface currents along the ocean floor. It is largely controlled by differences in water temperature and salinity.

total population growth how much the population of a place has grown, accounting for births, deaths, immigration and emigration

trade winds the easterly prevailing winds (north-east and south-east) blowing from sub-tropical areas towards the tropics

traditional society a labour-intensive agrarian economy and a relatively static population with limited technology

transmigration see **planned migration**

troposphere the lower 10–12 km of atmosphere where most weather activity takes place

tundra the very cold, flat and almost treeless biome below the northern Arctic. Because of the permafrost and snow cover for much of the year, there is only a short growing season for some mosses, lichens and grasses.

United Nations Climate Change Conferences (UNFCCC) annual meetings of heads of government to discuss strategies and progress relating to climate change targets and negotiate policies to reduce greenhouse emissions

urbanisation an increase in the total population or the percentage of a country's population living in urban areas

vegetation plant species and the ground cover provided by plants

voluntary migration when people migrate from one location to another by their own choice

Walker Circulation a model of air flow in the lower atmosphere across the oceans, consistent with differences in air temperature between continents and oceans. It is used as a measure for determining the onset of an El Niño event.

water cycle a model that demonstrates the movement of water on, above and below our environment

world fertility rate the number of children born per woman of childbearing age

yield gap the difference between current yield and yield potential

yield potential the output of a crop that can be achieved by regulation of genetic characteristics and temperature, etc

zainichi migrants or descendants of migrants from former Japanese-controlled territories in China and Korea

zooplankton the tiny marine creatures, such as krill, that consume phytoplankton and algae

Source: Geography 2019 v1.1 — General Senior Syllabus © State of Queensland (Queensland Curriculum & Assessment Authority) 2019; Various terms/definitions.

INDEX

Note: Figures and tables are indicated by italic *f* and *t*, respectively, following the page reference.

A

Aboriginal and Torres Strait Islander peoples 103–5
   agriculture 104–5
   fire 105
   housing and development 105
Africa
   desertification 48–9
   top five causes of death 204
age dependency ratio 204*f*
age of high mass consumption 196
age–sex patterns 148–53
age–sex pyramids 127
   for Maleny and Greater Brisbane 169*f*
   for the Rural North and Queensland 170*f*
age–sex structure 123, 193
   Malaysia 195*f*
age–standardised death rates 137, 137*f*
ageing population
   challenges of 203–4
   China 205–10
   Japan 207–10
ageing world
   ageing population *see* ageing population
   declining fertility 202–3
   increasing life expectancy 202–3
agriculture 86, 104–5
   crop yields 87*f*
   spatial pattern of 86–9
albedo 17–18
algal blooms 10, 101
Amazon, deforestation 40*f*
angiosperms 50
Antarctica, climate change on 68–70
anthropogenic activity and global warming 55–74
anthropogenic biomes 27–8
anthropogenic processes, in land cover change
   resource exploitation 78–9
   urbanisation 78
AO *see* Arctic Oscillation
aquaculture 51, 86
archipelagos 4
Arctic Oscillation (AO) 19–20

asylum seekers
   from Turkey to Greece 218*f*
   movement and destination 218*f*
atmosphere 9–12
   heat transfer 12–13
Australia
   demographic statistics for 158*t*
   fertility rate 141*f*
   health care 140*f*
   health landscape 139*f*
   overseas-born population 153*f*
   population density 121*f*
   population pyramid for 127*f*
   rainfall deficiency 96*f*
   uneven population distribution 147*f*
Australian Bureau of Statistics (ABS) 158–9

B

baby boom period 140
Bangladesh, climate change 72–3
barefoot doctors 205
biodiversity 29, 102
biogeographic areas 9, 21–2
biological weathering 81
biomes 9, 21–2
   types
      desertification 47–8
      deserts 44–7
      forests 29–33
      rainforests 30–3
      savanna and tropical grasslands 41–4
biosphere 9
birth rates 122, 140–2
   by country 124*f*
broadscale clearing 102
brown-field developments 173
Bureau of Meteorology (BOM) 19
Burgess' model 225, 226*f*
bushfire 82, 97*f*, 96–7

C

Cambodia
   forced internal migration 243–4
   population distribution 215*t*
   unequal distributions 212–16
canopy 32
carbon cycle 10–11

carbon dioxide (CO_2) 55
chaparrals 41
chemical weathering 81
Chicago, ethnic groups (1930), 226*t*
chilling effects 20*f*
China
   ageing population 205–10
   birth and death rates in 1950–2015 206*f*
   demographic transition 205–7
   dependency ratios 210*t*
   fertility rates 209*t*
   life expectancy (1950–2050) 205*f*
   one-child policy 206–7
   population growth 209*t*
   population pyramid 2017 208*f*
chlorofluorocarbons (CFC-11, CFC-12) 55
clearing of rainforests 33–41
climate change 80
   Antarctica 68–70
   in Bangladesh 72–3
   definition 3
   extreme weather events 3*f*
   greenhouse effect 18
   land cover change and 28–9
   mitigate 61–5
   in Queensland 61
Climate Change Conference 63
coal 95
coastal biogeographic areas 50–5
   coastal wetlands 50–1
   inland wetlands 51
   mangroves 50–5
   Ramsar sites 51–5
coastal land reclamation in China 52–3
coastal wetlands 50–1
coasts
   environmental impacts 100
   fieldwork ideas for 112
   land cover changes effects 100
composition 120
conflict 218–20
continents 4–5
Coriolis effect 12
correlation technique 236
crops 78
cumulative causation 196

curvilinear relationships   198
cyclone   82–3, 97–8, 98f

D

dam construction   23, 217
death rate   122, 192–3
declining mortality   192
deforestation   23
   effects of   34f
   global (1990–2015)   92f
   spatial pattern of   92–5
demographic challenges
   for inland towns   178–81
   for inner suburbs   171–5
   of suburban development on metropolitan fringes   175–7
   for tree- and sea-change towns   177–8
demographic concepts   122–7
demographic patterns of country towns, comparing   166–170
demographic patterns of suburbs, comparing   159–65
demographic transition   200–1
   China   205–10
   England   201f
   Japan   207–10
   Notestein's four stages of   200
   United Kingdom   200–2
   Wales   201f
demography   99, 120–2
dependency ratio   164
deposition   83–4
   of sediment   81–2
desertification   47–8
   Africa   48–9
   causes of   47f
   mitigate   48–9
deserts   44–7
   appearance and features   45
   human activities   46–7
   land cover changes   85–6
   location   45
   vegetation   46
   weather and climate   46
   wildlife   46
developed world   186
development   77
distribution   120
downwelling   14
drive to maturity   196
drought   83, 95–6
   escapers   46
   and food insecurity   220–1
   resisters   46
dunes systems   83–4

E

Earth
   affects by sun   13f
   continents   4–5
   covered by land   4
   energy budget   16
   oceans   4–5
   physical systems   9–12
Earth's physical systems connections   10–12
   carbon cycle   10–11
   hydrological cycle   10
   nitrogen cycle   11–12
ecological climax   30
economic development, population growth
   levels of   192
   Rostow's stages of growth   195–7, 196f
   theories of   195–200
   Wallerstein's world systems theory   197–200
economic factors of migration   221
economic growth
   Rostow's stages of   195–7, 196f
ecosystems   5
education factors of migration   222
EEC *see* European Economic Community (EEC)
El Niño phenomenon   15–16
El Niño Southern Oscillation (ENSO)   16
energy budget, Earth   16
energy flows
   in biosphere   80
   in ecosystems   81f
England
   demographic transition   201f
   ethnic groups   227f
   internal migration   240f
environmental factors of migration   222
Erg deserts   45
erosion   80–1, 83–4, *see also* soil erosion
   streambank   102
ESA *see* European Space Agency
estuarine ecosystems   10
ethnic enclaves
   and ghettos   225–6
   in UK   226–9
ethnic groups
   Chicago   226t
   England   227f
   largest population growth (2001–11)   228f
   Leicester (1991–2011)   227f
   Wales   227f
ethnic segregation   228
Eucalyptus forest   30f
Europe
   asylum seekers destination   218f
   population density   122f
   voluntary migration   230–3
European Economic Community (EEC)   222
European Space Agency (ESA)   6
European Union (EU)
   age structure of migrants   224f
   member states   231f
evapotranspiration   28
exposed soil   102

F

female workforce participation rates   142t
fertility rate   123, 142t, 189–192
   for Australia   141f
   by country   125f
fieldwork
   analyse the data   113–14
   for coastal areas   112
   communicate ideas   114–15
   evaluate the options   114
   gather information   111
   in local area   110–15
   plan   110–11
   propose action   114
   for rivers or other waterways   112
   for vegetated areas   113
   for urban areas   112
fire   105
Five-Year Plan, economic development   53
Food and Agricultural Organization (FAO)   30
food insecurity   211
forced migration   216–21
   Cambodia   243–4
   conflict   218–20
   dam construction   217
   drought and food insecurity   220–1
   natural hazards and disasters   216–17
   persecution   217–18
foreign labour, Subaru's supply chain   234f
forests   29–33, 84–5
   dominant land cover   29–33
   land cover changes   84–5
   effects   102–3

forests (cont.)
  recreation using 33
  types of 84
format options, fieldwork 111
fossil fuels 11, 78

G

gentrification 99
geographic inquiry model 110
geospatial technologies 113
ghettos 225–6
global birth and death rates (1960–2016) 131f
global climate systems 12–21
  Arctic Oscillation 19–20
  El Niño phenomenon 15–16
  greenhouse effect 18
  heat transfer 12–15
  Indian Ocean Dipole 18–19
  La Niña phenomenon 15–16
  ocean circulation 14–15
  polar vortex 20–1
  surface reflectivity, effects of 17–18
  wind patterns 12
Global Compact for Migration 223
global conferences, climate change and 63–5
global deforestation 92f
global population change 186–9
global population growth 130t
global warming 14, 55–74
  causes of 55–6
  effects of 56–8
  predictions 57f
global wind system 12f
GNI per capita ($US PPP) 123
governance, migration 222
grasslands 84–5
  land cover changes 84–5
Greater Capital City Statistical Areas (GCCSAs) 150t
greenfield developments 177
greenhouse effect 18
greenhouse gas emissions 63
greenhouse gases 18
  carbon dioxide (CO_2) 55
  causes of 56
  chlorofluorocarbons (CFC-11, CFC-12) 55
  methane (CH_4) 55
  nitrous oxide (N_2O) 55
  water vapour 55
gross domestic product (GDP)
  global pattern of 192f
gyres 14

H

Hadley Cells 13
hazards, spatial pattern of 95–8
health care
  advances in 138–40
  in Australia 140f
health landscape, Australia 139f
heat transfer 14–15
  in atmosphere 12–13
  ocean circulation 14–15
Holocene period 87
horse latitudes 12
housing and development 105
Humboldt Current 15
hydrological cycle 10
  see also water cycle
hydrosphere 9

I

IMF see International Monetary Fund (IMF)
import substitution policies 197
Indian Ocean Dipole (IOD) 18–19
  negative phase 19
  positive phase 19
Indigenous Australians 156–7
Indonesia, population density (2013) 242f
Industrial Revolution 11, 193
infant mortality 192–3
infectious diseases 137
inland wetlands 51
inorganic compound 10
Intergovernmental Panel on Climate Change (IPCC) 63
internal migration
  England 240f
  migration within countries 236–244
  UK 238–241
  in Vietnam 237–238
  Wales 240f
international migration 223
  forced migration 216–21
  migration impact 223–35
  people migrate 216–22
  voluntary migration 221–2
International Monetary Fund (IMF) 224
Intertropical Convergence Zone (ITCZ) 12
intra-state migration 143
IOD see Indian Ocean Dipole

IPCC see Intergovernmental Panel on Climate Change

J

Japan
  ageing population 207–10
  birth and death rates in (1950–2008) 207
  demographic transition 207–10
  dependency ratios 210t
  fertility rates 209t
  migration 234–5
  population growth 209
  population pyramid (2017) 208f
Japanese Ministry of Health, Labor and Welfare 207

K

Kyoto Protocol (1997) 63

L

La Niña phenomenon 15–16
land clearing 93–5
  dry country of Central Queensland 26
land cover
  biogeographical areas 21–2
  biomes 21–2
  patterns 5–9
land cover changes 21–9, 77, 106
  Aboriginal and Torres Strait Islander peoples 103–5
  Australia since European settlement 26–7
  climate change 28–9
  climate change and 80
  climate, impacts of 60–1
  coasts 83–4, 100
  connections between people and physical systems 103–5
  deserts 85–6
  effects of 99–103
  forests and grasslands 84
  forests and vegetated areas 108–9
  impacts in urban areas 99–100
  negative impacts 106–10
  rivers 108
  processes resulting in 78–86
    anthropogenic processes see anthropogenic processes
    natural processes see natural processes
  rivers 84
  spatial pattern of 86–99
    agriculture 86–9
    deforestation 92–5

hazards 95–9
mining 95
urbanisation 89–92
urban areas 83
least developed countries 124, 186
leeward side 45
Leicester, ethnic groups in 227f
life expectancy 123, 192–3, 194f
advances in 138–40
ageing world 202–3
at birth by country 124f
at birth (2016) 194f
in China (1950–2050) 205f
mortality 192–3
lithosphere 9
littoral drift 50
local area population patterns 158–70
comparing demographic patterns of country towns 166–70
of suburbs 159–65
local government area (LGA) 90
location quotient (LQ) 165
London
migration flows (2012) 241f
population growth (1931–2012) 239f
West Indian migrants (1950) 224f
longevity 130
Lorenz Curve 213
plotting data 214f
population distribution 213f
Lorenz, Max 213
luxuriant vegetation 32

M

Malaysia, age–sex structure 195f
male and female workforce participation rates 141f
Malthus, Thomas 193
mangrove root systems 51
median ages by state or territory 150t
methane (CH_4) 55
metropolitan areas, demographic change 171–7
migrant settlement patterns 153–6
migration
forced 216–21
impact of 223–35
in Japan 234–5
modelling patterns 235–6
north-west Europe 230f
planned 241–5
UK (1970–2013) 233f

voluntary 221–2
vs. human development index 236
within countries 236–45
migration policies over time 143–4
migration rates 123
over time 142–4
Millennium Development Goals 48
minerals 78
mining, spatial pattern of 95
mitigate climate change 61–5
mitigate desertification 48–9
mobility transition 235–6, 235f
Modern World System, The 197
Modernisation theory 195
monsoon systems 15
Montreal Protocol 55
more developed countries 124
mortality 192–3
mortality rates 123
multiplier effect 147
Murdie's model 225f
Murray–Darling Basin 101–2

N

NAO *see* North Atlantic Oscillation
national parks
using recreation forests 33
natural hazards
and disasters 216–17
and extreme weather 82–3
natural processes, in land cover change 80–3
deposition 81–2
energy flows in biosphere 80
hazards and extreme weather 82–3
weathering and erosion 80–1
net internal migration rate, Vietnam 237f
net interstate migration 142t
net migration 123
newly industrialised countries (NICs) 186
nitrogen cycle 11–12
nitrous oxide (N_2O) 55
nomadic herding 42
North Atlantic Oscillation (NAO) 19
nutrients 78

O

ocean circulation 14
oceans 4–5
human activities, effects of 5
Office of National Statistics (ONS) 240
one-child policy 206–7

ONS *see* Office of National Statistics (ONS)

P

patterns of land cover 5–9
people migrate
forced migration 216–21
reasons 216f
voluntary migration 221–2
percentage urban 123
permafrost 20
persecution 217
photosynthesis 11, 80
physical systems, earth 9–12
physical weathering 80
phytoplankton 11
planned migration 241–5
planning and structure, fieldwork 110–11
Pleistocene Ice Age 14
polar vortex 20–1
polders 72
Polish Resettlement Act of 1948 221
population
aged 65 and above by country 125f
patterns and trends 147–57
age–sex patterns 148–53
Indigenous Australians 156–7
migrant settlement patterns 153–6
population distribution 147–8
population census 186
population change 119, 130
demography 120–2
factors affecting 136–47
migration rates over time 142–4
natural increase 137–42
impact of disease on death rates 137–8
implications for people and places 170–82
metropolitan areas 171–7
regional and rural towns 177–82
in age–sex structure 134–6
local area population patterns 158–70
rates of natural increase and decrease 130–4
population density 120, *see also* population distribution
population distribution 147–8
Cambodia 215t
factors influencing 210–11

population distribution (*cont.*)
  impact of 211
  Lorenz Curve for 213*f*
  within countries 211–16
population growth
  by world regions 188*f*
  demographic components of 194*f*
  demographic transition model 200–2
  economic development 192, 195–200
  global population change 186–9
  population census 186
  rate 2015 188*f*
  selected countries and regions (1500–2000) 198*f*
  Uganda 190–1
  world and annual (1750–2050) 187*f*
  world population size and composition 189–95
population growth effects 27
population momentum growth 193
population pyramid 127
  for Australia 127*f*, 134*f*
  for Indigenous and non-Indigenous Australians 157*f*
  for Italy 128*f*
  for Niger 128*f*
  of West End 172*f*
population size 120
prairies 41
precipitation 10
preconditions to take-off 196
primary data 111
primary dune 83–4
protocols, climate change and 63–5
pull factors 143
push factors 143

Q

Queensland
  area by crop 88*t*

R

rain shadow 45
rainfall deficiency, Australia 96*f*
rainforests
  appearance and features 32
  clearing of 33–41
  human activities 33
  location 32
  vegetation 33
  wildlife 33
Ramsar Convention 51
Ramsar sites 51–5
rangelands 41–4
rate of natural increase 123
Red Zone, Christchurch 216–17
regional and rural towns, demographic change 177–82
remote sensing 6
replacement fertility 189
resource exploitation 78–9
resources 78
riparian zone 101
rivers
  fieldwork ideas for 112
  land cover changes effects 100–2
Rostow's stages of growth 195–7

S

salt intrusion 72
savanna *see* grasslands
savanna and grasslands 41–4
  appearance and features 42
  climate 42
  human activities 42–4
  location 41
  vegetation 42
  weather 42
  wildlife 42
  world coverage of 41*f*
savanna into farmland 44
seas 4
secondary data 111
secondary dunes 84
sequestration 11
sink 11
social factors of migration 222
socioeconomic status 99
soil erosion 81
soil management 81
soil, exposed 102
solar energy 80
soluble gas 11
Southern Ocean 4
Southern Oscillation Index (SOI) 16
spatial technologies 113
Spearman's rank correlation coefficient 145–7
standardised death rates 138*f*
steppes 41
streambank erosion 102
surface albedo 17
surface reflectivity effects of 17–18
sustainability 77
synoptic chart 61
Syrians
  movements of displaced (2017) 219*f*

T

take off 196
tertiary dunes 84
thermocline 16
thermohaline 14
Thompson, Warren 200
total population growth 123
trade winds 12
traditional society 196
transmigration 241
treaties, climate change and 63–5
tropical rainforests 31*f*
troposphere 9
tundra 21

U

Uganda
  population dynamics in 190*f*
  population growth 190–1
  population pyramid (2017) 91*f*
unemployment
  rates of (2015) 231, 232*t*
  United Kingdom 240*t*
United Kingdom (UK)
  demographic transition 200–2
  ethnic enclaves 226–30
  fiscal contribution of immigrants 233*f*
  internal migration 238–41
  migration (1970–2013) 226–30
  regional net migration 240*t*
  regions 239*f*
  unemployment 240*t*
United Nations (UN)
  census data 186
  population forecast (2017) 189*f*
United Nations (UN) 78
United Nations Climate Change Conferences (UNFCCC) 63
United Nations Population Fund (UNFPA) 191
United States (US)
  birth and death rates (1910–2010) 197*f*
urban areas
  fieldwork ideas for 112
  land cover changes 83
  impacts 99–100
urbanisation 78, 140
  spatial pattern of 89–92

V

vegetated areas
  fieldwork ideas for 113
vegetation 77

land cover change effects on 102–3
plant growth and 80
Vietnam
  internal migration 237–8
  net internal migration rate 237*f*
  types of migration 237*t*
voluntary migration 216, 221–2
  economic factors 221–2
  education factors 222
  environmental factors 222
  governance 222
  in Europe 230–4
  social factors 222

W

Wales
  demographic transition 201*f*
  ethnic groups 227*f*
  internal migration 240*f*
Walker Circulation 15
Wallerstein, Immanuel 197
Wallerstein's world systems theory 197–200, 198*f*
water cycle 84
  see also hydrological cycle
waterways
  fieldwork ideas for 112
  land cover changes effect on 100–2
weathering 80–1
West Bank
  planned Israeli settlement 244–5
West Papua
  transmigrants in 244
wildlife corridors 102
wind patterns 12
women in society, role of 140–2
World Bank 89
World Bank Groundswell 236
world fertility
  rate 189–92
world map
  distribution of 5*f*
world population 120*f*
  age–sex structure 193
  fertility rate 189
  future, size and composition 189–95
  mortality 192–3
world population pyramid 129*f*
World Wide Fund for Nature (WWF) 34, 92
world's forests, location of 29*f*

Y

yield gap 211
yield potential 211

Z

Zelinksy's model 235*f*
zooplankton 11